RN 43124 SCI £3

1,000,000 Books

are available to read at

Forgotten Books

www.ForgottenBooks.com

Read online
Download PDF
Purchase in print

ISBN 978-1-330-13552-5
PIBN 10034495

This book is a reproduction of an important historical work. Forgotten Books uses state-of-the-art technology to digitally reconstruct the work, preserving the original format whilst repairing imperfections present in the aged copy. In rare cases, an imperfection in the original, such as a blemish or missing page, may be replicated in our edition. We do, however, repair the vast majority of imperfections successfully; any imperfections that remain are intentionally left to preserve the state of such historical works.

Forgotten Books is a registered trademark of FB &c Ltd.
Copyright © 2018 FB &c Ltd.
FB &c Ltd, Dalton House, 60 Windsor Avenue, London, SW19 2RR.
Company number 08720141. Registered in England and Wales.

For support please visit www.forgottenbooks.com

1 MONTH OF FREE READING

at

www.ForgottenBooks.com

By purchasing this book you are eligible for one month membership to ForgottenBooks.com, giving you unlimited access to our entire collection of over 1,000,000 titles via our web site and mobile apps.

To claim your free month visit: www.forgottenbooks.com/free34495

* Offer is valid for 45 days from date of purchase. Terms and conditions apply.

English
Français
Deutsche
Italiano
Español
Português

www.forgottenbooks.com

Mythology Photography **Fiction**
Fishing Christianity **Art** Cooking
Essays Buddhism Freemasonry
Medicine **Biology** Music **Ancient Egypt** Evolution Carpentry Physics
Dance Geology **Mathematics** Fitness
Shakespeare **Folklore** Yoga Marketing
Confidence Immortality Biographies
Poetry **Psychology** Witchcraft
Electronics Chemistry History **Law**
Accounting **Philosophy** Anthropology
Alchemy Drama Quantum Mechanics
Atheism Sexual Health **Ancient History**
Entrepreneurship Languages Sport
Paleontology Needlework Islam
Metaphysics Investment Archaeology
Parenting Statistics Criminology
Motivational

HE HEAVENLY BODIE

THEIR

NATURE AND HABITABILITY

BY

WILLIAM MILLER, (S.S.C.)
of EDINBURGH
Author of 'Wintering in the Riviera'

LONDON
HODDER AND STOUGHTON
27 PATERNOSTER ROW
1883

100
n/a

PREFACE.

NEARLY five-and-thirty years ago an idea, entered in a note-book, occurred to me regarding 'plurality of worlds,' which at that time I imagined went far to unsettle the prevailing belief, and to decide a question subsequently much agitated. The little paper in which it was embodied, lying so long dormant, has been the germ of the present work. From a jotting on it I see I had in 1860 observed that another had, in a published book, struck out a similar idea, and it has since been repeated by other writers. It was then, if not previously, I presume, I became desirous to take up and examine the whole subject in all its bearings; but the incessant claims of an anxious profession altogether precluded the attempt, and it is only of recent date I have had it in my power to effect my purpose, and to bestow upon the subject the amount of study which I found to be requisite.

I may not, however, conceal from myself that many

will, with justice, consider that for one who cannot appear with the knowledge and authority of an astronomer, either practical or theoretical, it is a piece of bold presumption to endeavour to handle the matters to which this volume relates. In extenuation, I can only say that the occupations of my profession are favourable to the consideration of anything which demands close inquiry, and to the formation of soundly reasoned conclusions; while I have, in the course of an experience now extending over many years, had (sometimes in the conduct of what have been *causes célèbres* in their day) opportunities of studying, and generally of writing on, questions arising out of matters of extremely diversified nature. Nor have I dared to enter upon the scientific topics embraced in this book without investigating with, at least, the patient care I have always, when the rush and hurry of business would permit, sought to bring to bear on questions upon the solution of which material interests have depended. Still, in gathering, and dealing with, the evidence, which only others have been competent to furnish, I have necessarily experienced many difficulties, and would often have been glad to have had a chance of putting the witnesses under interrogation. I can only trust to have made unfamiliar subjects clear to the general apprehension, and that any opinions expressed will be found sufficiently supported. At same time, considering that these pages embrace so much of what to most people is new, it may not be unreasonable to expect, unless there have been failure to use the proper means for securing attention, that, if they do not allure to brighter worlds, they may at least help to lead the reader into fields of research which afford some of the purest and noblest subjects of meditation.

To discuss such a subject as the 'plurality of worlds' may be viewed differently by different minds. An American astronomer of the present day (Professor Newcomb) thus writes regarding it: 'The question whether other planets are, as a general rule, peopled, is one of the highest interest to us, not only as involving our place in creation, but as showing us what is really greatest in the Universe. Many thinking people regard the discovery of evidence of life in other worlds as the great ultimate object of telescopic research.' But without claiming for the subject such an undue importance, or availing myself of the time-honoured custom conceded to authors of 'magnifying mine office,' it can admit of no doubt that there is a certain amount of interest in the inquiry, and that it has awakened from time to time much attention; a fact which will be manifest from the historical *resumé* with which I have thought it well to commence. Perhaps the interest excited among the French may have exceeded that manifested by the more thoroughly practical people of this country. It will be seen how enthusiastically Fontenelle's work was received nigh two centuries ago, while to all appearance there has been among our philosophical neighbours no flagging of zealous regard for the subject, if one may judge from the fact of the very large number of editions through which Flammarion's *Mondes Habités* and *Mondes Imaginaires* have in a few years passed. And yet this French astronomer in contending for plurality seems to have contented himself as much as possible with the slender arguments previously advanced. The question, however, is one which, if it is to be settled on a solid basis, should be examined more searchingly, and my endeavour has been to consider it more fully and methodically. How far I have succeeded remains to be seen.

In pondering the 'plurality of worlds,' it speedily became apparent that a very wide range of subjects was involved, and that the idea which had occurred to me so many years before, although important, was in reality but a small point, going but a limited way towards the solution of this much vexed question. I must not, however, anticipate the conclusions to which examination has led.

The investigation occasioned divergence into other

PREFACE. ix

kindred matters which are not embraced in this volume, unless the chapters on the Sun are to be so regarded. But these chapters are truly german to the leading inquiry, which would hardly be complete without them. For it seemed advisable at the outset of examination to enter upon a careful analysis of what had been discovered regarding the nature and constitution of that important body, as well with a view to ascertain the possibility of its own habitability, as considering it in the light of a representative member of the starry system, from the grave bearing the ascertainment of the state of the fact must have upon conclusions to be drawn regarding the remaining stars. And of very recent years so much has been learnt with regard to our luminary, especially by means of that wonderful instrument the spectroscope, under the intelligent ingenuity with which it has been employed, that the knowledge now attained almost wholly upsets, or at least supersedes, what was formerly believed, and certainly furnishes us with a more exact acquaintance with a star so comparatively near. The material so acquired enabled me to deliver in Edinburgh two lectures on the Sun; and, while they embraced rather more than was strictly necessary to their due place in discussing the plurality of worlds, it has appeared to me that the facts stated are in themselves of much interest, and are more likely to be generally read when

PREFACE.

presented in this shape than if the reader had been referred to them as scattered over more scientific and formal works in which they might be found; but in the view of looking at the Sun as representing and illustrating the stars generally, it was well to carry with us a fuller notion of the leading circumstances connected with it, and therefore that they should be reproduced here, with such little additions and adaptations as seemed suitable, keeping always in view that the study of the Sun thus made was to lead up to the consideration of its condition as a residence for life.

I have indeed regarded it as a fortunate circumstance that at the time of the preparation of these Lectures I had seen no work affording, in a *systematic* manner, and up to date, the information which recent discoveries had brought to light; so that, although the investigation was thus rendered troublesome, it served to make it more thorough. Mr. Proctor's *Sun, Ruler, etc., of the Planetary System*, somehow I had not previously seen, and it was dated so far back as 1870, since which time a good deal has been revealed. Professor Young subsequently (1882) published in the International Scientific Series his interesting work, entitled *The Sun*, which, it may be well to keep in view, does not in all points accord with the opinions of Mr. Lockyer, by whose researches I had been largely guided.

In dealing with life upon the Earth, I have been led into a more extensive consideration of 'Evolution' than I had originally intended, or was perhaps required by its relative importance. But some of its advocates have sought to demonstrate their propositions and to overwhelm their antagonists by the force of dogmatic and rather arrogant assertion; so that it is well quietly to see, so far as needful for present purposes, whether their hypothesis can really be supported or no.

But upon this and other points I must leave the further discussion to more competent hands.

W. M.

GEORGE SQUARE,
EDINBURGH, *March* 1883.

CONTENTS.

	PAGE
PREFACE,	v

PART I.

THE SUN A REPRESENTATIVE STAR.

Introductory Observations,	1
1. The Sun's Magnitude, Distance, Motions, Light, and Heat,	5
2. The Constitution of the Sun, its Spots and Prominences,	41

PART II.

ARE THE HEAVENLY BODIES HABITABLE? OR, 'THE PLURALITY OF WORLDS' CONSIDERED.

Introductory Observations,	89
1. Historical *Resumé*—Views of Philosophers,	93
2. General Arguments for Plurality considered—Postulates,	133
3. The Sun, Stars, and supposed Stellar Systems,	145
(1) The Sun,	145
(2) The Fixed Stars,	151
(3) The supposed Stellar Planetary Systems,	157
4. Life upon the Earth,	171
5. The Moon,	218
6. The Minor Planets,	238
(1) Mercury,	238
(2) Venus,	244
(3) Mars,	249

		PAGE
7. Smaller Bodies of the System, ...		273
(1) The Asteroids and Meteorites, . . .		273
(2) The Comets, ...		282
8. The Major Planets, ...		300
(1) Jupiter, ...		300
(2) Saturn, ...		321
(3) Uranus, ...		332
(4) Neptune, ...		338
Conclusion, ...		341

PART I.

THE SUN A REPRESENTATIVE STAR.

THE SUN

A REPRESENTATIVE STAR.

THE solar system forms a minute part of the great galaxy called the Milky Way,—a galaxy covering so vast a portion of space that it has been said by a French writer that light, on the calculation, slightly erroneous, of its travelling at the rate of 186,000 miles a second, or upwards of 16,000 millions of miles a day, takes 15,000 years to pass from its one extremity to its other. The galaxy was estimated by Sir William Herschel to enclose in its main portion not less than 18 millions, and probably altogether will not number less than 20 millions of stars. And yet the Milky Way is only one of many, it may be thousands of, similar galaxies, which, separated from each other by tremendous gulfs, occupy the boundless universe, and are so remote that even their distant sheen is invisible to the naked

THE SUN.

eye, and only in part can be brought to sight by means of the most powerful telescopes.[1]

Some idea of the enormous number of stars thus filling the heavens may be got by a single computation. For were the 20 millions of stars of the Milky Way to be passed in review before us at the distance of the Sun (nearer would be undesirable, and even at so great a distance many would, from the potency of their rays of heat and light, be unendurable by us), and were they, large and small, to move on incessantly in close procession at the rate of only one per minute across the Sun's disc (equal to a possible average motion of nearly 60 millions of miles in the hour, or a million times quicker than the fastest railway train), it would take them about 37 years to march past.

Our Sun is also a star, as these stars are also all suns, and it is not a little strange to think we have a star

[1] Professor Newcomb is disposed only to allow from 30 to 50 millions of stars altogether. Arago, in place of reckoning, like Flammarion, 15,000 years for the Milky flight of light, puts it down at only 3000; while Dr. Dick, founding on Sir William Herschel's calculations, is content with 1640, and Lardner suggests 20,000. But astronomers disagree even more than doctors do, and in their estimates vary to an extent far exceeding the notorious divergences of competing builders. When one compares a dozen authorities on any point, it is hard to tell which to believe. Fortunately the difference of a few millions more or less does not materially affect either person or pocket, and equally fortunate is it that astronomical calculations are long, and life is short; so that astronomers are, occasionally, it is a pleasing relief to find, courteously inclined to borrow.

brought so comparatively near that it appears to us as a huge globe, and can now be made the subject of close ocular inspection. For, excepting the Moon, it is sometimes—or when Mars, Venus, and Mercury are not each approaching conjunction with the Earth—the nearest heavenly body.

The Sun therefore offers itself, and very steadily too, for our examination; and by aid of the means which inventive genius has, especially of late years, devised, examination has been rendered very complete. We have now possessed ourselves of a knowledge of its nature and characteristics, and of many facts previously unknown or obscure, so that the study of the bright luminary of day becomes profoundly interesting, and it is the more so, seeing that in giving a short account of the Sun, as it is now proposed to do, it will bring us by reason of their similarity into immediate acquaintance with the whole universe of stars.

In treating of the Sun it is convenient to deal with it in the two divisions which were originally the subject of separate discourse. The second division, in portraying its constitution, more thoroughly answers to inquiry as to the nature of the stars, the constitution of which is presumably similar. But the facts contained in the first division will materially contribute to the formation of a notion regarding other particulars which should go to complete our conjectural acquaintance with them. For in magnitude, motions, light, and heat there must be, with both individual and generic differences, undoubted general similarity; while the matter of distance is, in its vastness, a point equally applicable to their relations. The enormous immensity of the scale upon which everything connected with our luminary has been designed, is overwhelming to our feeble powers of conception. It will be the endeavour of these pages to make some effort towards realizing, although it can only be in a faint measure, what is so impossible for even the most vivid imagination fully or adequately to apprehend or to describe.

THE SUN'S MAGNITUDE, DISTANCE, MOTIONS, LIGHT, AND HEAT.[1]

By all the early pagans the Sun was regarded as a god, and was worshipped in different countries under different names, such as Baal, so repeatedly mentioned in sacred writ. Aristotle believed it to be animated. Some curious ideas were propounded by the Greeks. Thus Anaxagoras held the Sun to be a mass of ignited stone larger than the Peloponnesus, which, roughly speaking, may be said to be about 70 miles square, or somewhere about equal to the half of Scotland south of the Firth of Forth; and this minute fraction of the reality was doubtless regarded by the Greeks as indicating a heavenly body of immense size. Another, Anaximenes, maintained that, instead of being, as it appeared to be, a globe, it was flat or thin like a leaf; while Anaximander imagined it to be a chariot of fire, the fire escaping through a circular aperture or hole in the chariot. Zeno supposed the Sun to consist of a fire larger than the earth; while still another, Epicurus, represented the Sun as a terrestrial mass, pierced through like pumice-stone, and in a state of incandescence; also that the Sun was kindled in the morning, and was in the evening extinguished in the waters of the ocean, into which it

[1] Originally delivered at Edinburgh on 21st January 1882 as a lecture, which, however, has been here considerably altered.

seemed to the Greeks, from their geographical position, to fall.

Of a different kind, but perhaps to be put alongside of those dreams of unenlightened philosophers, was the opinion of an Englishman, the Rev. Tobias Swindon, who, so recently as 1727, published a book in which he set forth reasons for holding the Sun to be the local place of hell.[1] But a more common, though, as we shall speedily be convinced, a less justifiable idea, has, I suspect, been to consider it the abode of the blessed.

The invention of the telescope in the beginning of the 17th century put an end to such absurd notions as those to which the Greeks gave birth, and brought to man a more accurate knowledge of the heavenly bodies than he had previously possessed, but still led to great diversity and error of opinion regarding the nature of the Sun; and it is a circumstance not unworthy of remark, that Sir Isaac Newton, with the limited means of research which, in his time, were within command, but with the giant intellect he possessed, arrived at conclusions which, although only partially correct, have been considered to approximate those which astronomers of the present day, with vastly extended means of investigation, have at last come to hold. Before, however, entering upon a consideration of the constitution of the Sun, it is well to make ourselves somewhat acquainted with its size, its distance, its motions, its light, and its heat.

[1] This curious and rare old book was translated into French, *Recherches sur la nature de l'Enfer et du lieu où il est situé.* Traduit de l'Anglois par Jean Bron. Swindon, considering hell to be a place of real fire, held that, from its nature, size, position, etc., it is adapted to receive the fallen angels and the wicked. Only the French edition is to be found in the Advocates' Library.

THE MAGNITUDE OF THE SUN.

A good many years ago visitors to London seldom missed going to Leicester Square to see the great globe which was there erected as a model of the Earth. I suppose it was probably from 40 to 50 feet in diameter, or as large as a good-sized house. Upon it were delineated the features of the World—its oceans, mountains, rivers, and lands; and I confess that, small as this model comparatively was, it gave me, on walking round it, a better notion of the great planet on which we dwell than I had ever previously experienced. Now a model of the Sun on the same scale would need to be hewn out of a mountain a mile high and a mile thick, for it would require to be a mile in diameter. Placed beside the Leicester Square globe, we should obtain an idea of the comparative size of our great luminary.

Perhaps it may be well still further to gain an idea of the Sun's bulk by first trying to realize the size of this Earth. If we could soar on seraph's wings to a height of some hundreds of miles above the Earth, the grandeur of its proportions would vaguely fill our sight; but the altitudes within our reach by balloon or mountain ascent would give little notion; and we must seek it in a better way, and it may be had thus.[1] The circumference of the Earth is, we know, about 24,000 miles. Could we construct a continuous railway round it, and despatch over it an express train at ordinary express

[1] The greatest altitude ever attained in a balloon was on occasion of an ascent from Paris on 15th April 1875, when the aeronauts reached a height of 8 miles; but from the extreme rarity of the atmosphere at that great elevation two of them died, so that for scientific purposes it was recommended that thenceforth no ascent higher than 4 miles should be made. But these very lofty ascents by no means afford correspondingly good or satisfactory views. Mount Etna is 10,900 feet, or rather

speed, it would take, at 1000 miles a day, 24 days to make the circuit. That is, constantly travelling night and day by express train, we should be able to dash round the World in somewhat less than a month. This will give some conception of the magnitude of the World upon which we live. Its diameter is 7916, in round numbers 8000 miles; so that, if we could bore it and construct a tunnel through to the other side, and lay it with rails, we should emerge again to daylight at the antipodes,—say, in New Zealand,—by express, in about 8 days. The Earth, therefore, is in size by no means contemptible, and to us pigmy inhabitants it is very huge.

But the Sun is a body hitherto reckoned to be about 355,000 times the mass or weight of the Earth; while, seeing its substance is only about one-fourth of the Earth's density, its real size or volume is correspondingly greater, and is, as hitherto reckoned, 1,348,000 times the volume of the Earth. Now we can scarcely realize a body one thousand times greater than the Earth, but here we have one which is no less than nearly a million and a half times greater; although its distance is so vast that it appears to us as a comparatively small globe, no larger than the Moon, and it only looms large to us when we see it, as it rises or sets, resting on the top of a distant hill. The diameter of the Sun, however, in place of being, like the Earth, 8000 miles, is 882,646 (Young, 866,400[1]) miles, and its cir-

more than 2 miles high, and an extensive prospect must be had from its summit; but it only commands about 1/4000th part of the Earth's surface, which covers 200,000,000 square miles; so that to see the whole world we should require to have it studded at equal distances with 4000 mountains of similar height, and to ascend them all.

[1] Professor Young bases this and other relative calculations on the

ITS MAGNITUDE.

cumference is nearly 2,650,000 miles, or 111½ times that of our World. An express train, therefore, going 1000 miles a day, would take 7 years and 3 months to travel round the Sun.

Another method to adopt, is to think of the Sun as equal to 70,000,000 times that of the Moon. Conceive, if we can, what would be its apparent size were it brought as near to us as the Moon. Imagine a majestic body moving before us in the firmament in bulk 70,000,000 times, and in diameter 220 times, larger than the Moon.

These facts may give us some idea, but it may be realized in another way. Assuming the volume of the Sun at the above-mentioned 1,348,000 times that of the Earth, it would require so many globes each of the size of the Earth to make up a body equal to the Sun. Now suppose we were in imagination to take these 1,348,000 globes and put them in a line from the Sun out into space, they would extend to upwards of 10,600,000,000 miles. Now, as the Sun lies 93,000,000 miles from us, were these globes laid side by side in a straight line they would extend to about 114 times the distance we are from the Sun, or to nearly four times the distance from the Sun of the far-away planet Neptune, which, hugely greater than the Earth, is invisible to the unassisted eye,—a distance so great that an express train starting from the Sun would take upwards of 29,000 years to reach the last of the line of globes.

more recent estimates of the solar parallax. I have retained the previous estimates, and made my calculations on their basis, the actual size and distance of the Sun not being yet authoritatively ascertained. The difference is about 1/54th between the two sets of calculations, but this amount does not materially affect my illustrations.

It is only in some such way we can approximately realize the enormous mass of matter, or the vastness of the gigantic body of the Sun, which is more than 600, or, as Dr. Young states it, 750, times greater than that of all the planets, satellites, and comets of the solar system put together.

Indeed, so small is the Earth in comparison with the Sun, that Professor Tait says:—[1]

'The Earth, as seen from the Sun, appears very much less than the planet Jupiter, or even Mars, as seen by us,—that is, that it would present no visible disc to the naked eye, and that to an observer at such a distance as that of the Sun it would require a telescope of some little magnifying power to show it as a disc at all.'

On the other hand, we should run away with a very false notion if we supposed that the Sun is the largest body in space. On the contrary, it is believed on good ground to be a comparatively small star in the great universe; at all events, it is not among the largest, and the greater number of the stars, in all probability, are quite as large. Indeed, many of the fixed stars may be enormously larger. Mr. Proctor estimates Alpha Centauri, the nearest star, must be in diameter more than half as large again as the Sun, and its volume about five times his bulk.[2] Dr. Dick,[3] taking an observation of Sir William Herschel for his basis, calculates that the star Vega (a Lyræ), the third in distance from us (a brilliant star of the first magnitude in the constellation of the Harp), is about 38 times the diameter

---

[1] Tait's *Recent Advances in Physical Science*, p. 155.
[2] *Manchester Lectures*, 1872, p. 7.
[3] Dick's *Sidereal Heavens*, p. 64.

of the Sun, and that its solid contents are 54,872 times those of the Sun—a magnitude, he truly enough says, 'altogether overpowering to the human imagination.' Mr. Grant,[1] however, bringing out an even larger sum, attributes the result to an erroneous determination of the apparent diameter of the star. Yet it is probable that Vega is greatly larger than the Sun, though how much larger may be matter of opinion. But Sirius, which is farther removed than Vega, undoubtedly seems, by consensus of opinion, to be vastly larger than the Sun, although in this, as in so many other things, astronomers differ widely in their estimates. For Sir John Herschel estimates the volume of Sirius to be 8000 times that of the Sun; while Mr. Proctor, on what he conceives a better calculation of the distance of Sirius, though apparently by no means stating with any assurance, gives the bulk at only 4860 times— large enough certainly, but perhaps by no means approaching the size of some of the brilliant but vastly farther distant stars. Suppose, however, that either Vega or Sirius be only one thousand times larger than the Sun, their magnitude is beyond our conception; while, in all likelihood, there are stars much larger, perhaps, even than 54,000 times that of our Sun; stars so large as to exercise control over far-distant systems and worlds; magnitudes which cannot be even faintly apprehended by our finite powers. Such a star may be Alcyone, which is said, but not universally admitted, to be the central orb of a great star system of which the Sun and his planets form a part.[2]

[1] *History of Physical Astronomy*, p. 546.
[2] Newcomb, p. 208, states that for the most part the stars with a decided parallax are not of conspicuous magnitude, which indicates that those of conspicuous magnitude must be vastly larger.

Hardly anything can say, however little, of its size, be truly given in absolute figures.

DISTANCE OF THE SUN.

Here it is necessary to start with a knowledge of what the distance actually is in figures. It is a problem very far from being easy of solution.

"In the time of Copernicus it was supposed that the Sun's distance could not exceed 5 millions of miles, and indeed there are many who thought that estimate very extravagant. From a review of the observations of Tycho Brahe, Kepler, however, concluded the error was actually in the opposite direction, and that the estimate must be raised to at least 13 millions. In 1670, Cessini showed that these numbers were altogether inconsistent with the facts, and gave as his conclusion 85 millions."

Eighty-five millions was also erroneous, because short of the true distance. Fifty years ago children were taught at school that the distance was 95 millions. Now any schoolboy will probably affirm it to be 91 millions, but it turns out that every schoolboy is wrong, and that the real distance is different, and rather greater.

And this leads to the consideration of how the distance is determined. There are various methods, as is stated by Professor Forbes On the Transit of Venus, chaps. ii., also The Sun, by Dr. Young, p. 23. But the one which obtains most favour is that which arises out of the observation of the transits of the planet Venus, or its passing across the face of the Sun. These transits occur in pairs eight years apart, and with 122 years between the pairs. The last transit took place on 8th December 1874, and the next will be in December 1882;[1] and the next succeeding one will not be till 7th June 2004, that is, 122 years afterwards. Nor is a transit visible in all parts of the world. It is only seen—at least in its completeness—over a strip or belt of the world varying in position on each different occasion. It was first predicted by Kepler for the year 1631, but was not seen because that particular transit was not visible in Europe. A young English curate named Horrox, devoted to astronomical pursuits, thereafter calculated a transit for Sunday, 24th November 1639, the precise hour having been unknown, and there was even an uncertainty as to the day. After an anxious, protracted vigil,—interrupted on the Sunday by the requirements of clerical duty, so that the young enthusiast had alternately to rush off to service, doubtless hastily and abstractedly performed, and to rush back excitedly to observe and watch,—and nearly baffled by the occurrence of clouds, which, however, fortunately withdrew in sufficient time,—he was gratified, for the first time in the world's history, by observing the transit. His observations, though of a very rude kind, were important both in relation to Venus herself and as to the Sun's distance, although it does not appear that he himself observed with this view, and it is to Dr.

---

[1] This lecture was delivered previous to this the last transit, which, although the weather in Britain generally was unfavourable, has been seen elsewhere to great advantage, and excellent results from the observation are fully anticipated.

Halley we are indebted for the suggestion of making the transits a means of measuring the distance of the Sun. His dissertation on the subject, together with Ferguson's observations thereon, and various other relative papers, including account of Horrock's observation, will be found in Ferguson's *Astronomy*, pp. 434–501. Halley's suggestion was the more interesting that it bore reference to an event which would not happen for 70 years thereafter, and therefore long after he should be in his grave. The next transit took place in 1761, and the following one in 1769. On both these occasions several European States sent out expeditions to those parts of the world where observations could be made to most advantage. One of these expeditions was, under command of the famous Captain Cook, to the islands of the Pacific Ocean.

Great discussion followed on the observations then effected, showing much want of accordance in results— the computations of distance varying from 88 to 109 millions of miles, or no less than 21 millions of miles difference between the highest and the lowest. The celebrated mathematician Encke therefore devoted attention to the subject, and revised the calculations in 1822–24, announcing his result to be 95,274,000 miles; and this was held to be the true distance for a long time. Others subsequently computed the distance differently; but Leverrier at last made it 91,759,000, and Airy and Stone 91,400,000. 91 or $91\frac{1}{2}$ millions was therefore popularly assumed to be in round numbers the correct quantity. But astronomers were far from being satisfied, and waited impatiently for the next transit of Venus to enable them to make their calculation more exact. They had to wait a long time, for Venus, with feminine inflexibility, would not go out of her way one second to please the most impatient astronomer, charm he ever so wisely. The next transit did not happen till 1874.

By this time the means of observation had attained to vastly greater perfection than at the time of the previous transits, and all the civilised Powers, recognising the importance of ensuring absolute accuracy in observing the transit then to take place, provided expeditions to the different parts of the world where the transit would be seen. Books were published on the subject, including one by Mr. Proctor and a smaller one by Professor George Forbes. That by Mr. Proctor (who had previously published a series of papers on the subject in his book called *The Universe and the Coming Transits*) entered pretty fully into the history of the previous observations, and into a detail of the places and mode of observation, and is a model of beautiful illustration. That by Professor Forbes (more of a handbook), after explaining the object of the observation and the different means to be used for attaining results, so as to be checks on each other, gave an account of the various expeditions then to be undertaken by the different countries.

The British Government, besides making use of existing observations, selected special stations at Alexandria in Egypt, Kerguelen's Island (in the Southern Indian Ocean), the Sandwich Islands, Rodriquez, and New Zealand—two being situated in Kerguelen's Island, three in the Sandwich Islands, and two in Egypt, where there was also a private one at Thebes, by Colonel Campbell of Blythswood. The sum of £15,000 was originally voted by Parliament in aid; but how much more was subsequently voted I do not know. Over and above, Lord Lindsay undertook a station at the Mauritius, provided with appliances for utilizing all the different modes of observation, including about 50 chronometers; and Professor Forbes says it was perhaps the most completely equipped expedition which had ever been undertaken by a private individual in the interests of astronomy.

But without detailing the various other expeditions, I may simply say that the Germans sent out or stationed 5 or 6, the Russians 46, the French 8, the Americans 8, the Italians 3, the Dutch 1. In all, the transit was to be viewed from about 75 stations; and the expense of the whole was estimated to amount to between £150,000 and £200,000. The preparations were great for an observation which would scarcely last four hours, being the time Venus took to cross the Sun. But the observations then made were only the beginning of the inquiry. They afforded materials merely; for after they had been duly noted, they fell to be compared, and results deduced; and the number of calculations requiring to be made was so vast that it literally took years to overtake them. I have seen the number of calculations estimated, which was something astonishing, but unfortunately did not note it down at the time. It would have shown that the facts of astronomy are only obtained, not merely by means of much observation, both careful and minute, but by an after process of immense extent of calculation, and by the exercise of great skill, knowledge, and ingenuity, involving patient and laborious working, while they are very costly. Apart from their utility, they are of great interest to those who understand about them.

The result of all was to ascertain that previous calculations were in error. The Sun was considerably more distant from us than the last calculations had given. The British observers completed their calcula-

DETERMINATION OF DISTANCE. 17

tions, except what fell to be deduced from the photographic records, in the year 1876, and estimated the Sun's mean distance to be 93,300,000 miles, or nearly 2 millions more than the immediately previous estimate, and more than was anticipated; but I believe there has been as yet no final result arrived at by a collection and comparison of the observations and calculations of all countries.

Another method of determination has been recently adopted by observation of Mars when in opposition; and for this end Mr. Gill, Her Majesty's Astronomer, Cape of Good Hope, proceeded to the Island of Ascension in 1877. He has now published, in a reprint from the *Memoirs* of the Royal Astronomical Society, an account of his observations, with this

'Definitive result of the present investigation: mean horizonal equatorial parallax of the Sun = 8"·78, with the probable error ±0·012, which result combined with the most modern determination of the value of the equatorial radius of the Earth, viz. Listings = 3962·79 miles, gives for the mean distance of the Earth from the Sun 93,080,000 miles.'[1]

Dr. Young, referring to these labours, but before the issue of the final publication from which I have quoted, says:—

'So far as can be judged from the work thus far published, this determination must be conceded the precedence over all others in respect to its probable freedom from constant and systematic errors, and from theoretical difficulties.'[2]

Dr. Young, still[3] writing in the same predicament,

[1] *Memoirs*, p. 163.   [2] Young's *Sun*, p. 30.   [3] *Ibid.* p. 43.

'collecting all the evidence at present attainable,' arrives at a distance of 92,885,000 miles, with a probable error of 225,000 miles. Newcomb[1] estimates it to be between 92,200,000 and 92,700,000 miles.

The distance of the Earth from the Sun is not always the same, because its orbit is eliptical, not circular, and it also varies from year to year through a very long cycle or period of time, so that the difference between the two extremes is no less than 11,300,000 miles,[2] and the difference of heat at the two extremes is as 19 to 26.

'So slow, however,' says Ansted, 'is the change in eccentricity, that although it is now diminishing, and has been so for a long time, it will continue to do so for nearly 24,000 years without being reduced to a *minimum*. The actual distance of the Sun from the Earth changes every day; and owing to certain changes that take place in its movements, also occupying a long cycle (extending, however, over 25,686[3] years only), it happens that occasionally (as at present) the Earth is nearer to the Sun in the northern hemisphere during winter than during summer' by about 3 millions of miles. 'Nearly 10,000 years ago the Earth was nearest the Sun in summer, and farthest in winter, and the eccentricity was greater than it is now. This difference was certainly sufficient to produce a far more excessive climate, that is, a climate much hotter in summer and cooler in winter. As much as 210,000 years ago the difference was nearly a maximum in every way. Then perhaps was one of the glacial periods of geologists, for then the winter was nearly a month longer than the summer, as well as much colder than it is now.'[4]

[1] *Astronomy*, p. 200.   [2] Ansted's *Physical Geography*, p. 5.
[3] Sir J. Herschel has it 25,868 years, *Outlines*, § 316. Col. Drayson, 31,000 years, *Glacial Epoch*, p. 217.
[4] Ansted, p. 6.

REALIZING DISTANCE.

But assuming the mean or average distance of the Sun at the present time to be 93,000,000, let us endeavour to realize what this means. And in the first place we must start by seeing how long it takes to count up a million. Supposing, then, we could count 200 per minute, and did so for 10 hours a day, it would require above 8 days to count a million. To count up 93 millions would therefore occupy, at the same rate, no less than 2 years 3 months. From this we acquire some idea of what an enormous figure 93 millions comes to be.

We must, however, try and realize this 93 millions in miles.

Now the velocity at which light travels has been variously computed at different times, and is still subject of investigation. A considerable sum (£1000) was recently appropriated by the United States of America for the construction of apparatus to compute it. But the latest calculation is, I think, that by Dr. Huggins, who estimates that the rate at which light travels is 185,000 miles per second; and yet the distance is so great that light takes 8 minutes, or rather more, to reach us from the Sun, travelling at this inconceivable rate of speed. Yet, truly, this helps us but a very short way to realize or comprehend the immense distance; because we have no idea, and can form none, of a speed of 185,000 miles per second. Bearing in mind, however, that there are 480 seconds in 8 minutes, this enormous distance of 185,000 miles must be travelled 480 times over (it is really 500 times) before the light reaches the Earth.

Again, the Moon is distant from the Earth about 240,000 miles, and the Sun is removed from us about

390 times the distance of the Moon. Neither, however, can this calculation enable us to comprehend the vast distance—the immense gulf—which separates us from our great luminary.

As little can we comprehend the distance by being told that a cannon-ball discharged from a cannon would, it has been said, take 350 years to reach the Sun.

But suppose we adopt our former method of estimating —that is, according to time taken in railway travelling. Now, were a railway laid from the Earth to the Sun by which we could travel to it at the rate of 1000 miles per day, it would occupy no less than 255 years to perform the journey; so that had Charles I. on the day of his accession to the throne of Great Britain commenced a journey upon this imaginary railway, he could now only have just arrived at the other terminus, the Sun. Or if some adventurous hero were to begin it now, he could not reach the Sun till the year 2135; and he could not get back here, provided the Sun would let him, till the year 2390.

Travelling by special express train for 255 years would, however, be costly. Might we not try to walk the distance? Suppose, then, some remarkably good pedestrian, endowed with a body incapable of fatigue or decay, were to start on the journey and accomplish 30 miles a day, it would take him no less than 8415 years; and suppose he did enjoy the somewhat needful Sunday rest, he would not accomplish the journey in under 9617 years—that is, nearly 10,000 years, the odd 383 years being neither here nor there in an expedition of such duration.

Of course I am supposing a road laid down all the way as straight as a Roman road, or as many of the roads are in France to this day; and, like them, planted with an everlasting row of green shady trees on each side the highway, and supplied at least at the end of every day's journey with a comfortable hotel furnished with an abundance of the good things of this life, and conducted by a very obliging and accommodating landlord, willing to take drafts on the Sun in payment of his bill, and with here and there a tailor and shoemaker's shop to replace what was worn out by the way. The journey would indeed be a remarkably striking one,—perhaps *striking* in a double sense,—for the road would be rendered as lively as a city under bombardment, and a good deal more appalling, by the constant passage of huge flying bodies, and by coming in sight of terrific globes dashing nearer and nearer with furious and ever-increasing speed: of the Moon, of Venus, of Mercury whizzing with annihilating rush across the sky,—Venus and Mercury glowing with a scorching reflection of light and heat from the Sun; and comets flashing past with far-extending tail; and then as the brave traveller, who, we shall suppose, is protected from all evil consequences, came near the Sun itself, what an awful form it would assume, glowing as with millions of tremendous furnaces, and darting out its huge red flickering flames with a frightful rapidity, as if to lick up and suck everything in! He would, though scared and stupefied, be powerless to escape, and only be drawn on by an irresistible but horrible fascination, till unconsciously he touched the fiery willow leaves, the burning waves of the Sun, and should be at once ejected as an audacious, prying, interloping trespasser, and sent back on his weary course to Earth again, a wiser and a sadder man.

The thought of this dreadful walk of 10,000 years will give some idea of the prodigious distance away at which our immense luminary day after day rises in the east and turns his giant face upon us, benevolently beaming with light and heat and vivifying power. Were the Earth itself to lose that centrifugal force which keeps it from obeying the attraction of the Sun, and were thus, under the influence of the law of gravity, to be drawn to the parent orb, it would take little more than 60 days to perform its journey to the great parent orb; but long ere these two eventful months were over we should all have perished upon the face of the Earth; while the Earth, proceeding on with its freight of dead, would plunge into the Sun and be instantaneously swallowed up and lost, like a fragment of coal tossed into a huge smelting-furnace.

Yet, far as it is removed from the Earth, the Sun itself, with its surrounding planetary system, is still more inconceivably removed from even the nearest fixed star. The ascertainment, however, of the distances of even the nearest fixed stars is scarcely less difficult than the estimate of their magnitudes.[1] It has been, at least

[1] For the modes of measuring the distance of the stars, see Sir J. Herschel's *Outlines*, § 799 et seq. The Astronomer Royal thus informs us as to the method taken by Bessel to ascertain the distance of 61 Cygni, by observations commenced in August 1837 at Königsberg :—
'I mentioned another way in which the distances of the stars may be ascertained, a method which is free from all those defects of which I have spoken. This method is by the observation of two stars, of which one is believed to be very much nearer to the Sun than the other. For then we may assume that the distant star will have no sensible change of place from parallax, depending on the position of the Earth in its orbit. And then in observing the stars from the various parts of the Earth's orbit, we can compare the apparent place of that star which we believe to be the nearer with the place of the other. Practically this is of importance. The refraction, precession, mutation, and aberration are sensibly the same; and there is no uncertainty whatever from the computation of the various quantities which cast so much uncertainty on

approximately, made in the case of a few, the nearest of all being, as already mentioned, the star α Centauri. This star is calculated to be upwards of 200,000 times more distant than the Sun is from the Earth, its light taking 3 years 8 months to reach us. In miles the distance is 19,653,690,000,000 (tens of thousands of millions, more or less). But we can only faintly realize a figure so vast when it is added that an express train, travelling 1000 miles a day, would take upwards of 50 millions of years to reach this, our nearest starry neighbour. Yet the succeeding star is more than twice as far away.

The following table of distances has been furnished by Flammarion, corresponding with that of Guillemin :—[1]

	Number of Radii of Terrestrial Orbit.	Time Light travels to us.
α of the Centaur,	211,330	3 years 8 months.
α of the Swan,	550,920	9½ years.
Vega, α of Lyra,	1,330,700	21 ,,
Sirius, α of the Great Dog,	1,375,000	22 ,
α of the Great Bear,	1,550,800	25 ,.
Arcturus, α of Boötes,	1,622,800	26 ,
Pole Star,	3,078,600	50 ,,
The Goat, α of Auriga,	4,484,000	72 ,,

the results derived from other observations. This is the method pursued by Bessel in determining the distance of the star 61 Cygni. He measured the angular distance of this star from two small stars near it by means of an instrument called the heliometer, well known on the Continent, but of which there was at that time no specimen in England. With this he determined the parallax of the star 61 Cygni to be one-third of a second; that amounts to the same as saying that the distance is 600,000 times greater than the distance of the Earth from the Sun. It is deserving of attention that 61 Cygni is a double star; but we know from long observation that the two stars partake of the same motions, and probably are a connected system like the Earth and Moon, and therefore we speak of them and of their distance as if they were only one star.'—*Six Lectures on Astronomy at Ipswich*, by George Biddell Airy, Astronomer Royal, 4th ed. p. 170.

[1] Flammarion's *Wonders of Heavens*, p. 107; Guillemin's *Heavens*, p. 292.

These may not correspond with the figures of other astronomers, and indeed their calculations differ considerably, as may be seen by comparing with Newcomb;[1] but they suffice to give an idea of the enormous spaces by which one star is removed from another. There are stars in the Milky Way which are some thousand times as far removed as *a* Centauri; and Herschel believed his telescope revealed the light of stars which took 2 millions of years to reach us; so that, if he was right, these stars are 600,000 times as distant as the nearest star. One is overwhelmed with the thought.

But we require to take along with the distances the conception of

THE SUN'S MOTIONS.

The Sun does not in reality revolve on its axis so fast as the Earth, but its bulk is so greatly larger that the actual motion at the surface or circumference is nearly four times faster. Were the Earth to move at its surface as rapidly as the Sun moves at its surface, we should see the Sun, Moon, and stars perceptibly sailing through the heavens.

But besides its diurnal motion on its axis, the Sun is also moving through space, carrying along with it the Earth and all the other planets, and it does so with the prodigious velocity of well-nigh half a million of miles a day, or about 150,000,000 miles in a year. And

---

[1] Newcomb's *Astronomy*, p. 536. Mr. Proctor observes on the above list, that recent measurement reduces the distance of 61 Cygni and of Sirius; but that 'with the single exception, perhaps, of *a* Centauri, the stars are at distances beyond our present means of measurement.'—Guillemin's *Heavens*, p. 292, editorial note.

yet, at a speed which we cannot apprehend, it would occupy the Sun from 120,000 to 130,000 years to reach the nearest fixed star, were that star in its course, and would wait its arrival. The Sun is believed to be moving in a tremendous orbit of its own round some great central Sun, which, as already mentioned, has been rightly or wrongly supposed to be Alcyone in the Pleiades, dragging us all with it in its course. Yet, although moving at this inconceivable rate of 150,000,000 miles in a year, it is calculated to take 18,200,000 years to make the circuit, so that since the days of Adam it has only performed the 3000th part of its progress round that distant centre; and how often it has made the circuit no one can tell, or perhaps in this life ever will. Doubtless it has been many times. While all the other stars are in motion, too, revolving in great orbits of their own, never jarring or jostling each other, and, perhaps, all in turn are revolving round a great chief centre; yet in some cases these orbits, if they are orbits, must be so immense that the course of the stars appears to be in a straight line, and some are moving

'At such a rate that the gravitation of all the known stars cannot stop them until they shall have passed through and beyond the visible universe. The most remarkable of these, so far as we know, is Groombridge, 1830, it having the largest apparent motion of any star.'[1]

The motion of this star is so rapid that ' it would pass from the Earth to the Sun in about 5 days, so that its velocity probably exceeds 200 miles per second.' Some stars are receding from us, and others are approaching us (see a list in Newcomb, p. 458), but the star which is approaching us most directly is *a* Cygni, which is

[1] Newcomb's *Astronomy*, p. 485.

coming at the rate of about 40 miles per second, and in a period of between 100,000 and 300,000 years 'will pass by our system at something like a hundredth of its present distance, and will for several thousand years be many times nearer and brighter than any star is now.'[1]

These facts give rise to a curious thought. For, keeping in mind that every individual star is at a different distance from us,—some vastly farther off than others,—so that their light arrives at Earth in different periods of time, while each star is in rapid motion, and is moving in a separate course, the conclusion is forced upon us, that when we look upon the face of the sky we see not really a true picture of what at the exact moment surrounds us; for the light of one star having taken 1000 years to come to us, the star itself is now in a different place from where it was when its light left it; and the light of another having taken 5000 years, is still farther removed from where it was when the light left it, and thus what we actually see differs from what it would be did the light reach us instantaneously, and we then saw the stars just where they respectively are in the heavens.

Little as we usually think regarding the magnitude and distance of the Sun, I fear we as little apprehend, though we can more easily realize, its power and its importance to us.

'The Sun's rays,' says Sir John Herschel, 'are the ultimate source of almost every motion which takes place on the surface of the Earth. By its heat are produced all winds, and those disturbances in the electric equilibrium

[1] Newcomb's *Astronomy*, p. 459.

of the atmosphere which give rise to the phenomena of lightning, and probably also to those of terrestrial magnetism and the aurora. By their vivifying action vegetables are enabled to draw support from inorganic matter, and become in their turn the support of animals and of man, and the sources of those great deposits of dynamical efficiency which are laid up for human use in our coal strata. By them the waters of the sea are made to circulate in vapour through the air and irrigate the land, producing springs and rivers. By them are produced all disturbances of the chemical equilibrium of the elements of nature which, by a series of compositions and decompositions, give rise to new products, and originate a transfer of materials.'[1]

These, however, are only some of the uses of the Sun to us; but the fact is that all life upon the Earth is, under God, dependent upon the light and heat which we derive from it, so carefully adjusted and adapted that were they either considerably increased or considerably diminished, we should be either scorched or chilled out of existence, and the means of supporting life would likewise perish.

Let us, then, consider for a little the light and heat derived from the Sun.

THE SUN'S LIGHT.

And, first, with regard to that benign and glorious light with which the parent Sun gilds the Earth, cheering us from day to day, a blessing so freely bestowed that we think not of it except when it is cut off from us; and a mysterious thing it is, too, for it travels, as we have seen, at the enormous speed of 185,000 miles per second, and yet it is only a series of

[1] Herschel's *Outlines of Astronomy*, 8th ed. p. 259, § 399.

vibrations or waves 'dancing to and fro at the rate of about 450 billions of times in a single second,'—a rapidity of action which is truly as incomprehensible as are the figures by which it is expressed.

. The ascertainment of the intensity of the sunlight has long been a subject of experimental investigation, and among popular books will be found treated in Guillemin's *Sun*, chap. i., and Young's *Sun*, chap. viii., to which I would refer.

The comparison has been made chiefly with the light of candles. In 1725, Bouguer, a French astronomer and philosopher, made certain experiments, and taking his calculations as a basis, Guillemin estimates the light of the Sun at the zenith,—that is, when directly overhead,—with a clear sky, to be equal to 75,200 times that of a candle placed at the distance of $3\frac{1}{4}$ feet from the object illuminated. Dr. Wollaston, an English philosopher, made it to be rather less, or equal to about 68,000 times at similar distance. With such a basis of calculation the number of candles placed on a plane surface facing the Earth at the Sun's distance, so as to give a light equal to that of the Sun, has been computed, but is altogether beyond apprehension.[1]

M. Becquerel, another French natural philosopher, made experiments on the lime light, on the light arising from the burning of magnesium wire, and the electric light; and upon the footing of the estimate made of the power of candle light by Bouguer and Wollaston, Guillemin deduced that the Sun's light is equal to 75 times that of the electric light at $3\frac{1}{4}$ feet distance. With more powerful batteries other experimenters are said to have produced electric light than which the

[1] See Young's *Sun*, p. 240.

light of the Sun was only $2\frac{1}{2}$ times greater, which, if correct, was a near approach to daylight. It must, however, be borne in mind that the sunlight is most intense at noon, when the Sun is in the zenith. At sunrise it was found at Paris, on the 20th June (the height of summer), the light was 1300 times less than when the Sun was in the zenith, and that as the Sun rose higher and higher in the heavens the light proportionally got stronger and stronger.

The light of the Sun has also been compared with that of the Moon and of the stars.

Bouguer concluded the light of the Sun to be 300,000 times that of the Moon; Wollaston, as much as 801,072 times; Professor Bond of Cambridge, 470,000 times; Zöllner, 619,000 times as much light as the full Moon,—a result which Newcomb thinks is probably quite near the truth.

Huygens, the celebrated Dutch philosopher of the 17th century, estimated the light of the Sun to be equal to 765 millions of times that of Sirius, the most brilliant star in the sky; though in our hazy climate we do not see it glowing with the intensity with which it may be seen in the clearer southern atmosphere of the Riviera. But Wollaston's calculation, made at a long after period, and, doubtless, with much better means, was much greater, for he estimated the light of the Sun to be 20,000 millions of times that of Sirius;[1] and yet if Sirius were unfortunately brought as near to us as is the Sun, its light would be equal to 94 Suns similar to our own—a blaze of light by which we should immediately be blinded. The distance of Sirius from

---

[1] Great reliance on these computations, however, in the nature of the case, cannot be placed. Sir John Herschel and Mr. Bond estimated the light of Sirius at less.

us is, however, so enormously great,—seven times that of the nearest star,—that, assuming Wollaston's calculation to be correct, it would require the sky to be set— and it would be a gorgeous illumination—with 20,000 millions of visible stars, each with all the brilliance of Sirius, to light us up as fully as we are lighted by the Sun.

These facts show that there are stars which yield a much greater light than the Sun. Sirius is a particularly bright star. Sir John Herschel has estimated the intensity of its light to be equal to the light of nearly 200 stars of the 6th magnitude. But it is comparatively near, and some of the brilliant stars which are much farther removed may shine with a far more exceeding splendour.

THE SUN'S HEAT.

From the light of the Sun it is an easy transition to pass to consider its heat, for all light produces or carries with it heat. Dr. Lardner, in his treatise on heat,[1] does indeed say that moonlight is an exception; because in whatever degree moonlight can be concentrated by the most powerful burning glasses, it 'has never yet been found to affect the most sensible thermometer.' But his book is dated 1833, and, doubtless, expressed the result of experiments made during upwards of 130 years previously. Subsequent inquiries upset the view. For Professor Piazzi Smith[2] (who experimented on the top of the Peak of Teneriffe), Professor Forbes, and Melloni, an Italian, all have detected, though in minute degree, some amount

[1] Lardner's *Heat*, p. 351.   [2] See Proctor's *Moon*, pp. 272-282.

of heat in the Moon's rays; while Lord Rosse, using a powerful reflecting telescope, was able to say that the light reflected from the Moon can be felt, and he estimated that this heat is the 1/900,000th part of that radiated from the Sun.[1] The light emanating from all the stars, or from space, has been calculated by M. Guillemin[2] (and the statement is surprising) to afford the Earth as much heat as is equal to about 5/6th of the heat we derive from the Sun. This estimate is founded upon investigations by M. Pouillet, another Frenchman (for it is chiefly to French philosophers we are indebted for investigations on this subject), who found the temperature of interplanetary space to be about 140° C. below freezing point—a statement which it seems difficult to reconcile with Guillemin's calculation. The heat received from individual stars has likewise been estimated, and necessarily is minute; that from Arcturus was found to be 'equal to the radiation of a Leslie cube of boiling water at a distance of 383 yards.'[3]

But we do not by a long way receive from the Sun all the heat it emanates. Professor Tait tells us 'that the quantity of heat which the whole Earth gets from the Sun is of the order of something less than the 2000 millionth part of that which the Sun gives out.'[4]

Guillemin states the same thing: 'the heat intercepted by the Earth is only $\frac{1}{2,150,000,000}$ part of the entire solar radiation.'[5] Further, Sir John Herschel informs us that the heat received by the Earth in a given area is as

---

[1] Proctor's *Orbs around us*, p. 73.   [2] Guillemin's *Sun*, p. 37.
[3] Guillemin's *Forces of Nature*, p. 496.   [4] Tait's *Advances*, p. 155.
[5] Guillemin's *Forces of Nature*, p. 495.

to the heat on an equal given area in the Sun only about the 92,000th part,[1] or as he has elsewhere said, I presume meaning much the same thing, 'the temperature, that is to say, the degree or intensity of the heat, at the actual surface of the Sun' is 'more than 90,000 times greater than the intensity of sunshine here on our globe at noon and under the equator.'[2]

Putting these facts or calculations together, the amazing amount of heat which exists in, and is constantly proceeding from the Sun, must be something inconceivable; and from early times philosophers have set to work to estimate its intensity and measure its power. Recently M. Pouillet has invented an instrument (and others have done the same) to measure the solar heat as received by the Earth, called the Pyrheliometer.

In one of the most recent works on astronomy,—though in parts somewhat sketchy,—that by M. Rambosson, Laureate of the Institute of France,[3] there are given some of the different opinions which have been held on the subject of the Sun's temperature, from which it appears that so great is the disagreement of philosophers, that the heat of the Sun has been variously estimated at from 1461° up to 10,000,000°, the highest calculation being 6000 times that of the lowest, from which it is very evident that no great reliance can be placed upon any such estimates, although the progress of science tends to greater certainty. How-

---

[1] *Outlines*, § 396.
[2] Quoted in Williams' *Fuel of Sun*, p. 73, without reference.
[3] Rambosson's *Astronomy*, p. 93.

ever, even the lowest of these computations states an amount of heat far beyond our experience.[1] Endeavour has indeed been made to realize the amount of heat by comparing it with temperatures which can be reached on Earth. Thus it has been estimated by one man that the heat *emitted* from the Sun (which is greatly less than that contained in it) is 1000° hotter than what is necessary to melt iron; and Sir William Thomson has said it is from 15 to 45 times greater than that developed in the furnace of a locomotive, while he considers the Sun's radiation to be equivalent to about 7000 horse-power per square foot of his surface.[2] While such figures may be mentioned, all we can understand is, that the heat is something inconceivably tremendous.

The more common methods of computation or comparison, however, are by estimating by its power on ice,—by the amount of coal it would be necessary to burn to make its equivalent,—by its capacity for boiling, and the like.

Computations as with ice seem to be a favourite method. For example, Sir John Herschel[3] considers the heat, if it could be concentrated, sufficient to melt a cylinder of ice 45 miles in diameter continuously darted into the Sun with the velocity of light without diminishing the intensity of the Sun's heat. [4]Dr.

---

[1] 'Secchi originally contended for a temperature of about 18,000,000° Fahr. (though he afterwards lowered his estimate to about 250,000°). Ericson puts the figure at 4,000,000° or 5,000,000°. Zöllner, Spoerer, and Lane name temperatures ranging from 50,000° to 100,000° Fahr.; while Pouillet, Vicaire, and Deville have put it as low as between 3000° and 10,000° Fahr. The intensest artificial heat may perhaps reach 4000° Fahr.'—Young's *Sun*, p. 265.

[2] Tait's *Advances*, p. 158.   [3] *Outlines*, p. 259.   [4] Young's *Sun*, p. 255.

Young says it would in a second of time melt a column of ice $2\frac{1}{4}$ miles in diameter spanning the inconceivable abyss between the Earth and the Sun, and in seven seconds it would dissipate the whole in vapour. Pouillet[1] describes the heat as sufficient to melt in one minute a layer of ice 11.80 metres (nearly 40 feet) thickness, or in one day 16,992 metres, or $4\frac{1}{4}$ leagues,—that is, were the Sun suddenly girt with ice of adequate depth, the power of its heat would in one day melt through a belt of ice 13 miles thick.

Professor Tyndall[2] estimates this heat as equal to boiling 2900 thousand millions of cubic kilometres, or as it is elsewhere stated, 700,000,000 cubic miles of water at the temperature of ice. Tyndall[3] also says: 'Expressed in another way, the heat emitted by the Sun is equal to that which would be produced by the combustion of a seam of coal 27 kilometres (16.65 miles) thick.' Sir John Herschel puts it, that to maintain the heat calculated by Sir William Thomson, it would require 13,500 lbs. for every square yard of the Sun—that is, 6 tons of coal to be burnt each hour on each square yard of the Sun.

But I only mention these estimates, which really it is impossible for us to realize, in order to furnish some idea, vague and general it must be, of the vast heat existing in and emanating from the Sun even after allowing the largest deduction for over-estimates. It is an amount (whichever calculation we choose) with which we have nothing on Earth fortunately to compare; for 'if the Sun were to come as near us as the Moon, the solid Earth would melt like wax.'[4] The experiments

---

[1] Williams' *Fuel of Sun*, p. 74. [2] Guillemin's *Sun*, p. 35.
[3] Guillemin's *Forces of Nature*, p. 495. [4] Young's *Sun*, p. 268.

HEAT—HOW PRODUCED. 35

and observations made by some of the French philosophers[1] led them to conclude that the solar heat does not exceed 3000° centigrade, and probably is from 2500° to 2800°, in other words 12 or 14 times the heat necessary to make water boil. These are lower estimates than others, and seem to be unreliable. Rosetti finds, says Dr. Young—

'18,000° Fahr. as the *effective temperature* of the Sun,—a result which, all things considered, seems to the writer more reasonable and better founded than any of the earlier estimates. Rosetti considers that this is also pretty nearly the actual temperature of the upper layers of the photosphere."[2]

But really, after all, no more can be said on this subject than that philosophers widely disagree as to the extent or degree of heat, but all agree that it is something far exceeding anything we have experience of here, or which we can possibly realize.

I do not enter upon the power of the Sun's heat further than to mention that when the Sun's rays are collected and concentrated by means of a burning glass, the heat exceeds that of a powerful furnace, and is sufficient to melt gold and convert it into vapour.[3] It is also a curious fact to be stated in the same connection, that the heat may be transmitted through a burning glass of ice, so as, without melting the ice itself, to set on fire what is placed in the focus.

How this awful heat is produced or maintained is a question which has not yet been solved. Sir John Herschel says:—

---

[1] Rambosson, p. 95. [2] Young's *Sun*, p. 267.
[3] Lardner's *Heat*, p. 350.

'The great mystery is to conceive how so enormous a conflagration (if such it be) can be kept up. Every discovery in chemical science here leaves us completely at a loss, or rather seems to remove further the prospect of probable explanation.'[1]

'It is quite certain,' says Dr. Young, 'that it is not a case of mere combustion.' 'If the Sun were made of solid coal burning in pure oxygen, it could only last about six thousand years.[2]

Various theories have been propounded, and one of them, by M. Matthieu Williams in *Fuel of the Sun*, who, discarding a supposition that it is maintained by a bombardment of aerolites, suggests the ingenious explanation that the Sun, travelling through space with the velocity of about 400 to 500 thousand miles a day, is fed by the air through which it passes, which he calculates would be equal to fresh fuel at the rate of 165 millions of millions of tons per second. In other words, that the Sun is an immense fire fed by fuel. But if I might hazard an observation, I think this view is untenable, because—1st, It would imply the Sun is fed only upon the side or part or half which is in advance,—the other half or back part would be unfed, or at least meagrely fed while in rear. And as the Sun takes nearly a month to perform its revolution on its axis, each part of its surface would in succession be unfed for nearly half a month. 2d, Its condition would necessarily be dependent upon the quality of the atmosphere or space through which it passed,—in some places densely laden with matter, in others lightly, so that it would burn unsteadily. 3d, The probability is that the matter of the system moves with the Sun or system, so that its material

[1] *Outlines*, § 400.   [2] Young, p. 270.

would soon be burnt out. 4th, That no appearances in the shape of a trail of smoke or otherwise justify the conclusion. There may be better scientific reason, and indeed any form of maintenance by combustion seems now to be given up, and the view generally entertained appears rather to be that first suggested by Helmholtz, that the great heat is engendered by the tremendous pressure and contraction which the force of gravity should produce; and this also is the view taken by Dr. Young, the latest authority, who, however, allows that a fraction of the heat may result from the falling in of meteors.

It is not doubted that the heat of the Sun is diminishing;[1] but it is comforting to think that, according to Father Secchi of Rome, the diminution 'is almost imperceptible, not exceeding one degree in 4000 years.' The rate, however, is a point upon which it is not possible to form any correct opinion. Professor Tait says: 'The Sun cannot possibly cool more than a single degree centigrade in seven years. It may be on the highest estimate we can take—one degree in 7000 years; the data are very uncertain; but we may say that these are the limits between which it must lie.'[2] But taking the highest estimate of 7000, and the temperature of the Sun to be now 18,000° (and it may be vastly more), the conclusion would be that in 126,000,000 years the sun will cool down to zero. Dr. Young, saying that Newcomb allows it only 10,000,000 years to continue to afford sufficient heat to support life, will not himself allow the possibility of 60 millions of years.[3] Indeed, he would assign a total existence for the solar system from its birth to its death

---

[1] See Professor Nichol's *Phenomena*, p. 191.
[2] *Recent Advances*, p. 159. [3] Young's *Sun*, p. 276.

of only 30 millions of years,[1]—a fraction merely of the periods which others have assigned.

The Sun's heat, however, might be maintained for a short season longer than it would otherwise be by the falling into it of the planets—a supposition not impossible. For from time to time (though so rarely that, as according to Humboldt, only twenty have been recorded as seen in 2000 years) there have been instances of stars observed suddenly to blaze out as if on fire,—one of the most remarkable of which was that of a star in the constellation Corona borealis, which, on 12th May 1866, burst out for a period of twelve days from the appearance of a star of the 9th magnitude to that of the 2d, and then returned to its former magnitude;[2] and there have not been found wanting those who have expressed the opinion that such a blazing out might be occasioned by the violent precipitation of some great mass, perhaps of a planet upon the star. Indeed, it is now considered, although the period it may take may be very long protracted, that the existence of aluminiferous[3] ether in the realms of space, united to other causes, must offer such a resistance to the planets[4] as will in course of time cause them to fall into the Sun; just as the aerolites or falling stars, flying through space, are arrested by the Earth at the rate of seven and a half millions per day.[5]

---

[1] Young's *Sun*, p. 277.

[2] Schellen's *Spectrum Analysis*, p. 523. See also mention of the star in Cassiopeia, and the star η Argus in Professor Nichol's *Solar System*, p. 189.

[3] Helmholtz, *Popular Scientific Lectures*, 2d series, p. 165.

[4] The resistance experienced in the case of Encke's comet has, however, it is right to say, been attributed by some to its passing through the zodiacal nebulosity.

[5] Helmholtz, p. 169.

Professor Tait[1] calculates that if the Earth were to fall into the Sun, the heat thus generated would be equal to the heat given out by the Sun in about 91 years. Sir William Thomson[2] makes it 95 years, and calculates the additional duration which the collapse of the other planets would ensure. Jupiter falling in would, from its immensely greater bulk, add 32,254 years, while all the planets together would only effect a prolongation of the Sun's beams for 45,604 years. But after all, the Sun must in time be reduced to a black body like that of the companion of Sirius, which is almost indiscernable from the faintness of its light,[3] and ultimately to that of the satellites or companion of the star Procyon, which, from its darkness, has not yet been seen.

Mr. Kalley Miller, contemplating speculations of this nature, observes:[4] 'When the last of the planets is swallowed up, the Sun's energies will rapidly die out, and a deep and deathly gloom gather around nature's grave. Looking into the ages of a future eternity, we can see nothing but a cold and burnt-out mass remaining of that glorious orb which went forth in the morning of time, joyful as a bridegroom from his chamber, and rejoicing as a strong man to run a race.'

[1] *Recent Advances*, p. 158.   [2] Young, p. 272.
[3] Helmholtz, p. 190.
[4] R. Kalley Miller's *Romance of Astronomy*, p. 108.

II.

THE CONSTITUTION OF THE SUN, ITS SPOTS AND PROMINENCES.[1]

PRINCIPAL LEITCH of Queen's College, Canada, in a popular work called *God's Glory in the Heavens*, published in 1863 (eighteen years ago), expressing an idea which at that time and for ninety years previously had prevailed among astronomers, thus describes our subject:[2]—

'The Sun may then be conceived as composed of a dark central body, encompassed by successive envelopes or shells suspended at different heights in the atmosphere, the uppermost being the one which forms the luminous disc of the Sun. A Chinese ivory ball, composed of carved concentric shells, represents very well the structure of the Sun and the nature of the spots.'

Upon this dark central body the supposed inhabitants of the Sun were imagined to dwell. Imprisoned within this tremendous oven, enveloped by a vast surrounding canopy of cloud and fire which would for ever exclude the view of all outside the Sun, its poor denizens, we might fairly conceive, if not undergoing the daily torment of continuous suffocation, would, without some

---

[1] This lecture was delivered before the Literary Institute of Edinburgh on 7th December 1881, and was repeated, at request, before the Greenock Philosophical Society on 24th February 1882. A few alterations and additions have been made since.

[2] 2d edition, p. 123.

marvellous means of refrigeration, be unceasingly suffering the horrid pangs of roasting or stewing alive.

We are, however, saved the pain of realizing and sympathizing with so frightful a condition by the knowledge to which we have now attained. During these eighteen years astronomers, with the enormously increased means of investigation at their disposal, and especially by the use of the spectroscope, have been vigorously prosecuting their inquiries—with gigantic strides have succeeded in unveiling many mysteries, and have intelligently and patiently achieved that apparently difficult task of throwing light upon the Sun. Nay, we are not content with having our eyes filled with the Sun, but we must have our ears too. For within the last few months attempt, it seems, has been made by Professor Graham Bell, under the guidance of M. Janssen, a famous French astronomer, by employment of the photophone, and that with sufficient encouragement, it is said, to justify hope of ultimate success, to explore the noises supposed to accompany the vast movements of matter taking place in the solar surface.

Perhaps, in Great Britain at least, Mr. Norman Lockyer is the astronomer who has, by his unwearied labours, most largely contributed to acquaintance with our great luminary. Among the foreign astronomers, we are indebted to the late Father Secchi of Rome, M. Janssen, and others, some of whose names I shall have occasion to mention.

Philosophers, owing to the limited means of observation formerly had, have, till recent years, widely differed regarding the constitution of the Sun. Some maintained, like Professor Leitch, that its interior is, or contains, a more or less solid body,—cool and habitable; others that the Sun is entirely gaseous. The former opinion is now abandoned, and the latter proposition, that it is gaseous, is what is now generally held as the result of more accurate scientific investigation to be its true explanation or description.

This, then, is now considered to be the general formation of the Sun—

(1.) The interior or portion which lies below or within what is called the photosphere or surface, and is entirely invisible to us.

(2.) The *photosphere*, or luminous face of the Sun.

(3.) Above this the *chromosphere*, or denser portion of the Sun's atmosphere, and

(4.) Above the chromosphere, extending to a great height, the *Corona*, or thinner or more attenuated atmosphere.

Let us consider these different divisions in their order.

1. *The Interior of the Sun.*

Though invisible, there is every reason to believe it to be composed of gaseous vapours in a state of incandescence, burning with inconceivable heat; and this conclusion is supported by the well-ascertained fact that the density of the Sun is so much less than that of the Earth as to be only one and a half times that of water.

'So that,' says Sir John Herschel,[1] ' it must consist in reality of far *lighter* materials, especially when we consider the force under which its central parts must be

[1] *Outlines*, 8th ed. § 449, p. 297.

condensed. This consideration renders it highly probable that intense heat prevails in its interior, by which its elasticity is reinforced and rendered capable of resisting this almost inconceivable pressure without collapsing into smaller dimensions.'

M. Faye, a French astronomer—

'Regards the interior of the Sun as consisting of the original nebula from which our whole system has been slowly condensed, in a state of dissociation—that is, at such an intense heat that chemical combinations are impossible.'[1]

Father Secchi, late Director of the Observatory at Rome, who has largely contributed to our knowledge of the Sun by his observations on this, to him, a favourite subject, says:—

'When the sun at the epoch of its formation had reached a volume about equivalent to that which it now possesses, its temperature would have been at least 500 millions degrees; and, moreover, we know by experiments that even now its surface temperature amounts to several millions degrees; that of the interior is probably higher still. We must conclude from these facts that the Sun cannot be composed of a solid mass; nor, enormous as may be the pressure existent in this mass, it cannot possibly, so to speak, be in a liquid state. Whence we are necessarily led to the supposition that it is gaseous notwithstanding its extreme condensation.'[2]

M. Delauney, another French astronomer, holds that—

'The Sun is a gaseous mass with a very elevated temperature, which prevents the elementary substances that enter into its composition from consolidating.'[3]

[1] Lockyer's *Solar Physics*, p. 67.   [2] Rambosson's *Astronomy*, p. 89.
[3] The views of some of the distinguished students of the Sun, including specially Secchi, Faye, Young, and Langley, will be found in Newcomb's *Astronomy*, p. 265 et seq.

M. Rambosson, Laureate of the Institute of France, says:—

'For my own part, after comparing the various solutions that have been proposed, I must pronounce for the gaseous nature of the sun.'[1]

Professor Young, holding it probable that the Sun's core is gaseous, observes:—

'Nothing could be remoter from the truth than to imagine that a mass of gas under such conditions of temperature and pressure' (as he had described) 'would resemble our air in its obvious characteristics. It would be denser than water; and since, as Maxwell and others have shown, the viscosity of a gas increases fast with rising temperature, it is probable that it would resist motion something like a mass of pitch or putty.'[2]

Such is the concurrence of opinion; and the fact appears to be pretty universally admitted or proved, that the heat in its intensity is in the centre,—Mr. Lockyer observing that there is proof that the deeper we go the hotter we get,—and the coolness, comparative of course, is to be found in the exterior or crust. It is to this comparative coolness—this which to us would be an inappreciable diminution of inconceivably fervent heat—the Sun owes

2. *The Photosphere, its shining cover.*
M. Faye, Mr. Lockyer observes—

'Looks upon the photosphere as the surface at which the heat is so acted upon by the cold of space as to allow chemical combinations and solid and liquid particles to exist.'[3]

[1] P. 90.   [2] *Sun*, p. 286.   [3] *Solar Physics*, p. 67.

It is from this photosphere, or shining skin of the Sun, we derive our light and our heat, and its appearance when examined by the telescope is very remarkable. Observation of it had been made, as stated by Arago,[1] from time to time. It had been remarked by the Bavarian Jesuit Scheiner, a contemporary of Galileo, that the entire surface of the solar body was constantly covered with light and dark streaks of slender dimensions. A century and a half later (1774) Francis Wollaston, an English astronomer, said, 'The Sun is frequently dotted, perhaps always is,' and Sir William Herschel in 1795 wrote, 'The Sun appears to be irregular, like the skin of an orange.' About twenty years ago Mr. Nasmyth, an English astronomer, examining with more care, observed upon the Sun what, from their shape and elongated form, he designated, 'Willow leaves,' a name by which this peculiar mottling has generally been since known. It has been likened, however, by other observers to slashed blades of straw, and to grains of rice. Although only visible under a powerful telescope, these so-called willow leaves are necessarily of great size, having been estimated to be not less than 1000 miles long by from 200 to 300 miles broad. They have been said by some, though it is doubted or disputed by others, to interlace each other like the cross hatching in a drawing or engraving. The leaves themselves are luminous. The spaces between them, being by contrast dark or less luminous, have been called pores or openings. The leaves have thus been described by Sir John Herschel:—

'According to his (Mr. Nasmyth's) observations made with a very fine telescope of his own making, the bright surface of the Sun consists of separate insulated in-

[1] Arago's *Popular Astronomy*, p. 440.

dividual objects or things all nearly, or exactly, of one certain definite size and shape, which is more like that of a willow leaf, as he describes them, than anything else. These leaves or scales are not arranged in any order (as those on a butterfly's wing are), but lie crossing one another in all directions (like what are called spills in the game of spillikens), except at the borders of a spot where they point, for the most part, inwards towards the middle of the spot, presenting much the sort of appearance that the small leaves of some water-plants or sea-weeds do at the edge of a deep hole of clear water. The exceedingly definite shape of these objects, their exact similarity one to another, and the way in which they lie across and athwart each other (except where they form a sort of bridge across a spot, in which case they seem to affect a common direction, that, namely, of the bridge itself), all these characters seem quite repugnant to the notion of their being of a vaporous, a cloudy, or a fluid nature. Nothing remains but to consider them as separate and independent sheets, flakes, or scales having some sort of solidity. And these flakes, be they what they may, and whatever may be said about the dashing of meteoric stones into the Sun's atmosphere, etc., are evidently the *immediate sources of the solar light and heat ;*' and Sir John makes this singular addition: 'We cannot refuse to regard them as organisms of some peculiar and amazing kind, and though it would be too daring to speak of such organization as partaking of the nature of life, yet we do know that vital action is competent to develope both heat, light, and electricity.'[1]

The views relative to the supposed willow leaves have, however, now undergone a change. Mr. Lockyer, in his recent work on star-gazing (published 1878),[2] concludes, as the result of the work of many careful

---

[1] *Good Words*, 1863, p. 282.
[2] Lockyer's *Star-gazing*, p. 472. See also Skertchly's *Physical System of the Universe*, p. 222; Williams' *Fuel of the Sun*, p. 99.

observers since Mr. Nasmyth's discovery, that the appearance of mottling is due to dome-like masses; and these dome-like masses are not improbably produced by the upward and downward currents, to which I shall afterwards refer.

Indeed, it has been a great subject of controversy whether this surface or photosphere is solid, liquid, or gaseous. The result of the most recent investigations seems to be the establishment of the position that it is gaseous. Nor do I think this is at all inconsistent with the idea of a certain cloudy and even firm consistency or crystallization, as it were, into shapes more or less resembling the form of willow leaves. Our own clouds, which are not subjected to the same enormous pressure, often assume forms having an apparent solidity or compactness dependent for their shapes upon the currents existing in the atmosphere in which they are suspended. But the surface of the Sun is subject to the pressure of an atmosphere which, we shall presently see, is several hundred thousand miles high, and this pressure, combined with the contraction occasioned to a metallic gas by exposure to a certain amount of (comparatively speaking) cooler influence, may cause the luminous matter to assume given forms, such as that of the willow leaf.

There are, however, breaks in the willow-leaf system; for by means of the telescope it is seen that there are here and there upon it, but constantly varying in number, extent, and place, what have been termed 'spots.' I shall deal with them more specially afterwards. Meantime, I may just say generally that these are huge cavities or holes or irregular wells in the photosphere.

On the other hand, there are raised portions which may be considered to be mountains of light, and are called *faculæ*. These mountains or ridges always exist round the edges or borders of a spot, but are found all over the Sun, while the spots are not. They form the most intensely brilliant parts, and may be—

'Of all magnitudes,' says Mr. Lockyer, 'from hardly visible, softly gleaming, narrow tracts 1000 miles long, to continuous, complicated, and heapy ridges 40,000 miles and more in length, and 1000 to 4000 miles broad.'[1]

The height of these ridges no one, so far as I have noticed, seems to have estimated, unless it be Dr. Dick,[2] who speaks of them as several hundred miles above the level; but they would require to be of greater height to be even seen as ridges. They have been attributed to uprushes from the interior of the Sun; and if so, may be somewhat akin to the uprush of molten lava from a volcano, although probably not possessed of the same degree of viscous solidity. They have been observed to undergo a melting or disappearing action, and under the influences existing on the Sun's surface in time may form portions of the willow leaf system, thus becoming feeders of it. This curious operation is described by Mr. Lockyer,[3] who, in observing 'a tongue of facula' stretching half-way into a spot, at first extremely brilliant, but afterwards losing this brilliancy, says:—

'At the same time it seemed to me to be "giving out," as it were, at its end, and a portion of the umbra between it and the penumbra appeared to be veiled with a stratus cloud evolved out of it.[4] After a time, large, very dim

---

[1] P. 18.   [2] Dick's *Celestial Scenery*, ed. 1860, p. 215.
[3] Lockyer's *Solar Physics*, p. 26.
[4] I have seen at Zermatt an appearance analogous to this, in the issue from the Matterhorn, under the influence of a strong Sun melting the snow, of a continuous streak of white cloud.

"willow leaves" seemed to be forming (condensing) on the following portion of the cloudy mass. So that at first you got a very brilliant mass of what appeared to be facula gradually melting away into umbra, and then the umbra condensing into willow leaves.'

3. *The Chromosphere.*

Surrounding the photosphere like an envelope, the spectroscope has revealed the existence of what was considered by Arago and other astronomers to be clouds, and which some thought shone by reflected light. Mr. Lockyer, who was the first, or an early observer of certain lines produced by it on the spectroscope, says[1] wherever he looked he found these newly discovered lines—

'Showing that for some 5000 miles in height all round the Sun there was an envelope, of which the prominences were but the higher waves. This envelope I named the chromosphere, as it is the region in which all the variously coloured effects are seen in total eclipses, and because I considered it of importance to distinguish between its discontinuous spectrum and the continuous one of the photosphere.'

Apparently this envelope 'as a rule bounds the convection currents of the Sun,'[2] and does not exactly or at all places rest on the photosphere or surface of the Sun,[3] but lies, like our clouds, a little above it, leaving hollows between, which are, Mr. Lockyer says, doubtless filled with dense vapours. Father Secchi fancies, apparently, that the space of separation extends to as much as 40,000 miles.[4] But this view does not seem to be concurred in by other astronomers, and is opposed by Mr. Lockyer. The chromosphere itself, though largely

---

[1] Lockyer, p. 218.   [2] *Ibid.* p. 417.
[3] *Ibid.* p. 409.   [4] Schellen, *Spectrum Analysis*, p. 414.

consisting of hydrogen gas, is pervaded by the vapours of various metals.[1] It is not of uniform thickness, and its thickness seems to depend on the prominences or eruptions from below, to which I shall afterwards advert.

4. *The Corona.*

Above the chromosphere is a luminous atmosphere, only visible during total eclipses, called *the Corona*. It is thus described by Mr. Lockyer:—

'A total eclipse of the Sun is at once one of the grandest and most awe-inspiring sights it is possible for man to witness. All nature conspires to make it strange and unearthly. Soon the stars burst out, and surrounding the dark Moon on all sides is seen a glorious halo, generally of a silver white light; this is called the Corona. It is slightly radiated in structure, and extends sometimes beyond the Moon to a distance equal to her diameter. Besides this, rays of light, called aigrettes, diverge from the Moon's edge, and appear to be shining through the light of the corona. In some eclipses parts of the corona have reached to a much greater distance from the Moon's edge than in others. It is supposed that the corona is the Sun's atmosphere, which is not seen when the Sun itself is visible, owing to the overpowering light of the latter.'[2]

But the speculation regarding the corona has been great. Up to the year 1842 it was firmly believed to be an appendage of the Moon,[3] and indicated the height of its supposed atmosphere,—an opinion which was, Mr. Lockyer mentions, not banished from men's minds till 1860.[4] Thereafter the idea prevailed, and apparently was at one time entertained by Mr. Lockyer himself, that it owed its origin to the Earth's atmosphere.

[1] Lockyer, p. 480.
[2] *Ibid.* p. 74.
[3] *Ibid.* p. 106.
[4] *Ibid.* pp. 254, 265, 271.

The subject strongly interested astronomers, and became a special question for determination by observation in succeeding eclipses, in which all the aids to consideration were had recourse to, particularly during the eclipses of 1869, 1870, and 1871, specially mentioned in Mr. Lockyer's *Solar Physics*, published in 1874, and also in subsequent eclipses. The methods taken embraced drawings of the appearances as seen by the naked eye, spectroscopic and polariscopic observations, and photography. The observations by these different methods differed, but the representations obtained by means of photography during the eclipse of 1871[1] seemed to establish beyond any doubt 'the solar nature of most, if not all, of the corona recorded on the plates.'

The eclipse of 1875 established this conclusively.[2] Dr. Janssen, well known for his previous observations, watched it carefully, and ascertained that the corona was in reality an atmosphere of the Sun radiating by itself.

A curious appearance,—perhaps due to an ocular deception,—but observed by different persons on different occasions, has been not merely that of a flickering or wavering, but even of a rotatory motion—the corona seeming to move in every conceivable way and direction. One man in 1788 said 'it seemed to be endued with a rapid rotatory motion, which caused it to resemble a firework turning round its centre.' Another in 1860 similarly described it as 'a thing that was going round and round like a firework.'[3] It is, however, very clear that the corona assumes different shapes and appearances and magnitudes at different times, and especially in

---

[1] Lockyer, p. 376.   [2] *Year Book of Facts in Science and Arts*, 1876, p. 248.
[3] Lockyer, pp. 294, 377.

being at times broken here and there by rifts. One cannot wonder, therefore, that some people have suggested 'that the corona was nothing but a permanent solar aurora.'[1]

It has, however, now been authoritatively established that this corona is in reality an attenuated atmosphere of the Sun;[2] but opinions have greatly differed as to the height above to which it extends, the most extravagant opinion ascribing to it 'two or three millions of miles in height.' Opinions have also differed and fluctuated from time to time in regard to the composition and pressure of the atmosphere.

The colour or appearance of the corona has been variously described. Dr. Lardner, referring to the observations of Mr. Hind and Mr. Dawes, says:—

'Its colour seemed to be that of tarnished silver, brightest next the Moon's limb, and gradually fading to a distance equal to one-third of her diameter, where it became confounded with the general tint of the heavens.'[3]

Guillemin describes it as 'sometimes of a pearly or silvery white, sometimes yellowish, and even red.' Father Fauro, writing to Father Secchi, says:—

'The colour was beyond the power of any artist to paint. All observers agree that it resembled mother-of-pearl, or pure unpolished silver, but far more beautiful, and more intensely brilliant.'[4]

The opinion entertained not many years ago was

[1] Lockyer, p. 305.
[2] Mr. Proctor, who gives much interesting information regarding the corona, seems to attribute it to an enormous mass of meteors accumulated round the Sun, I presume in a dissolved state (*Sun*, p. 364 et seq.).
[3] Lardner's *Handbook of Astronomy*, § 522.
[4] Schellen, *Spectrum Analysis*, § 325.

that the chromosphere is low and of simple composition, subject to a small pressure, and that the envelope of the Sun's atmosphere consists solely of hydrogen. All this is now changed, and there has been a discovery of two new substances in the solar atmosphere. One of these was at one time thought to be vapour of iron, but this opinion was abandoned, and the view was retained that it is the vapour or gas of a substance unknown on the Earth—a gas which seems to pervade the Sun's atmosphere from the lowest depths to the greatest elevation. It is denoted by the line in the spectrum, No. 1474. The other substance discovered is the gas of an also unknown constituent, and designated as the line D_3. It exists only at the lower altitudes. Both these gases, however, have recently been thought by Mr. Proctor to be a form of hydrogen, though differing from the ordinary forms of that gas as known to us.[1] But this is only a supposition which remains to be, if possible, proved. Hydrogen proper was formerly supposed to be the lightest of all gases or substances, but the unknown gas, 1474, is discovered existing still lighter, for it rises high above the utmost limit of hydrogen in the solar atmosphere, and forms the outer envelope, extending to at least 20', or about 500,000 miles high, and probably, could it be detected, and without doubt occasionally, a good deal higher; but at a height of 20',[2] ten thousand times higher than the atmosphere of the Earth, which has commonly been believed to extend only to a height of 50 miles, although a much greater elevation has been assigned to it by some. The line of extreme limit of hydrogen (proper) in the Sun's atmosphere is thought to be 12', or about 300,000 miles high, but at this height it

---

[1] *Year Book*, 1879, p. 78.   [2] 1' is = 25,000 miles upon the Sun.

exists as a layer of cool gas; below this again is a layer of incandescent hydrogen gas (that is gas at a white heat), mixed with vapours of various metals and of other substances. The hydrogen rests on a layer of the unknown element (D_3), and it again on the chromosphere, which extends 5000 miles above the Sun, and is of a composite character; while the chromosphere, in its under surface more or less pervaded by hollows,[1] rests separated probably by some space on the photosphere, and the photosphere reposes (if such a term can be applied to a scene of constant fiery activity) on the gaseous matter below,—thus floating like a crust on the atmosphere of the Sun, by which on both sides it is enveloped. Mr. Lockyer describes this succession of elements on the Sun in these words:—

'Travelling down from x', which gives us the 1474 line, and exists at the extremest, the most utterly distant parts of the corona, right down through the solar atmosphere to the bottom of the deepest spot, we shall pass very much through the different substances in this order. Beginning with the 1474 element, we pass through the sub-incandescent hydrogen; deeper still we get to the incandescent hydrogen; then we go through the D_3 element; then we get into regions where the lines are generally mixed rather more together, but from which magnesium and sodium are generally ejected more frequently and higher than any other material; then we get into the more doubtful zone of barium and nickel—sometimes sodium being thrown up, sometimes barium, sometimes nickel; and then we come lower down into what may be called, so far as we shall ever be able to investigate the Sun, the very bowels of our central orb, where we are certain to get iron, and we may get many other materials.'[2]

Although the corona extends to high above the Sun,

[1] Schellen, p. 415. [2] Lockyer, p. 415.

it must by no means be thought that it resembles in its upper portions the atmosphere of our Earth in density. The corona, it is said, must be hundreds of times lighter than hydrogen. The great comet of 1843 passed at the rate of 350 miles per second through at least 300,000 miles of it without visible damage or retardation. Whereas shooting stars are instantly and completely vaporized by encountering our atmosphere at a height of 50 or 100 miles.[1]

The existence and non-existence of substances in the Sun, which are known to us on Earth, have been discovered by means of the spectroscope, and different observers have made different lists or tables; but all are agreed that hydrogen, sodium, iron, calcium, magnesium, and nickel are there. As to other substances, opinions differ; but there is, at all events, a general accord that gold, silver, mercury, tin, and lead, —all metals having a great specific gravity,—with some other substances existing in the Earth, are wanting. What have been so discovered all exist, in consequence of the frightful heat, in a state of vapour. The absence of the heavy metals may, I think, be explained on the supposition, according to the Nebular hypothesis, that they may find their place in the outer planets.

Recently oxygen has been discovered by Professor Henry Draper.[2] The existence of nitrogen is not yet with certainty ascertained.[3] Professor Newcomb regards the discovery of oxygen as 'the most important advance in spectrum analysis since Lockyer and Janssen discovered the spectrum of the solar protuberances.'[4]

[1] Newcomb, p. 259.
[2] See Guillemin's *Forces of Nature*, Appendix, p. 673.
[3] *Ibid.* p. 676.   [4] *Ibid.* p. 520.

With this enormous atmosphere, laden, too, with so many metallic vapours, joined to the immense attraction or force of gravity exerted by the huge body of the Sun, the pressure on the photosphere must be excessive, and to that enormous pressure may be due, as experiment would seem to prove, the brilliancy of the Sun's light;[1] while below the photosphere the pressure must maintain, if it do not produce, the inconceivable heat which fills the vast gaseous cavity forming the interior of the Sun.

We are now perhaps in a position to approach the consideration of the spots and prominences on the Sun, both of which, like everything else connected with the investigation of our great luminary, have, by reason of diversity of opinion, been the subjects of much discussion.

THE SOLAR SPOTS.

The spots on the Sun are not usually visible to the naked or unassisted eye. It is to Galileo, the wonderful Italian, the discovery of them is commonly, and apparently with truth, attributed. M. Arago,[2] no doubt, disputes this position, and dates observation back to the time of Virgil, citing cases of observations from the year 321 downwards to 1607 of spots which, on some occasions, were supposed to have been the planets Mercury or Venus crossing the Sun's disc. Such observations without telescopic aid, however, can hardly be regarded as having been of much value, and the first real observation was truly made by Galileo who, in 1610, soon after he had invented his telescope, and even with the very small power he had at first succeeded in producing,

[1] Lockyer, p. 329. [2] Book XIV. c. viii. p. 420.

observed, and informed others of, the spots on the Sun's disc, and from their motion inferred, as one named John Fabricius is said by Arago[1] to have first done, the revolution of the Sun upon its axis. The honour of the telescopic discovery was indeed contested by the Jesuit Father Scheiner, professor of mathematics at Ingolstadt; but he did not in reality observe them for a year later. Galileo, in a work published by him in March 1613, called *History and Explanation of the Solar Spots*,[2] settled the controversy between him and Scheiner and his other opponents; while he at the same time first published his views in favour of the Copernican system—a publication which soon brought him into trouble with the ecclesiastical powers of Italy.

Galileo's opinion with regard to the spots was that they are matter in or near the Sun of the nature of cloud, and are not permanent; that they are black or dark to appearance in the centre, but are in reality as bright as the Moon. Scheiner, on the other hand, held them to be similar to planetary moons or comets, and that some might be 'as far from the Sun as the Moon, Venus, or Mercury (on the Ptolemaic system).'[3] They are not, he thought, *on* the Sun, but, although thin, were as dense as the Moon.

Towards the close of the century, Fontenelle, in his once popular but superficial work on the *Plurality of Worlds*, states, and seems to countenance, the absurd notion that the Sun is of melted gold, which appears to boil over continually, and by force of its motion casts up the spots as scum or dross to its surface, where it is consumed.[4]

[1] *Popular Astronomy*, p. 424.
[2] Gebler's *Galileo*, p. 44.
[3] Lockyer, p. 8.
[4] Fontenelle, p. 104.

NATURE OF SPOTS.

Dr. Wilson of Glasgow, in 1774, showed that the spots are cavities,[1] or what he terms 'excavations,' in a luminous envelope surrounding the Sun, which, according to his view, was a dark globe. Surrounding the dark spot there always is a lighter portion, which has been called the penumbra, and Wilson inferred that the penumbræ were the sloping sides of the well at the bottom of which is the black spot.[2] The justice of which deduction, Skertchly says,[3] concurred in by Guillemin,[4] 'all subsequent research has confirmed.' In 1769, Wilson also calculated, according to mathematical principles, that the depth of one spot was equal to the Earth's radius, or 4000 miles;[5] but generally he assigned a depth of from 2000 to 4000 miles.[6] Matthieu Williams, however, considers 'one or two hundred thousand miles'[7] as not an unreasonable estimate of the depth of some of the largest Sun spots. But this is likely to be as extravagant an estimate as the other is inadequate.

A very common opinion, held up till recent times, as we have seen, and one conceived by Dr. Wilson, and shared in by Sir William Herschel, and more recently by the late Professor Nichol of Glasgow,[8] and others, was that the Sun is a solid body, and the spots are cavities or displacements in the shining envelope, through which we see this solid body or nucleus, the cavities being occasioned by the working of some sort of elastic vapour which is generated within the dark globe. The French philosopher, La Lande, suggested that the spots

---

[1] See Nichol's *Phenomena of Solar System*, p. 178 et seq.; also his *Solar System*, p. 170.
[2] Arago, p. 436.
[3] Skertchly, p. 207.
[4] Guillemin's *Sun*, p. 222.
[5] Arago, p. 438.
[6] Dick's *Celestial Scenery*, p. 213.
[7] Williams' *Fuel of Sun*, p. 90.
[8] Nichol's *Phenomena*, pp. 186–190.

might be 'eminences in the nature of mountains actually laid bare.'[1] The idea that a spot was disclosure of the assumed solid body of the Sun, was rather countenanced by the observations of Mr. Dawes in 1851, who supposed he had found the blacker portions of spots merely to be an inferior or lower—

'Stratum of very feebly luminous or illuminated matter, which he has called the cloudy stratum, which again in its turn is frequently seen to be pierced with a smaller and usually much more rounded aperture, which would seem at length to afford a view of the real solar surface of most intense blackness.'[2]

All these and other opinions were formed during the time when the Sun could only be examined by means of the telescope, which, even using the finest instrument 'under the most favourable atmospheric conditions,' only enables us to observe the various phenomena as we should do with the naked eye at a distance of 180,000 miles. But the invention of the spectroscope, among its other wonders, has at last, after patient investigation and consideration, and not without the broaching of discordant theories, settled, according to Lockyer, that the spots are occasioned by a downrush from the upper and cooler solar atmosphere of the metallic vapours[3] with which it is loaded through the photosphere to the hotter regions below. A spot, accordingly, is just, in the words of Mr. Lockyer—

'A place in which principally the vapours of sodium, barium, iron, and magnesium occupy a lower level than they do ordinarily in the atmosphere.' It is 'the seat of a downrush or downsinking' (sometimes having the ap-

---

[1] Sir J. Herschel's *Outlines*, § 389.   [2] *Ibid.* § 389 *a*.
[3] Lockyer, pp. 70, 72.

pearance of being formed in a whirlpool[1]), and the greater darkness of a spot is due to 'the general absorption of the atmosphere, thicker here than elsewhere, as the spot is a cavity.'[2]

This result of investigation has accounted also for the fact of the gradations of darkness observed in spots, the darker portions being those which are deepest, and probably most densely laden with metallic fumes.

Such is the simple explanation of solar spots. But there are various circumstances connected with them.

1. In the first place, a spot is often of *tremendous size.* Guillemin says:—

'In 1779, William Herschel saw a spot which was not less than 50,000 miles in diameter. Those shown in fig. 30, from a drawing by Captain Davis, show what enormous dimensions some spots sometimes attain, the largest of them, which has, however, a double nucleus, was not less than 187,000 miles in its greatest length; its superfices, including the penumbra, was about 25,000,000,000 square miles. If these spots are, as we shall see farther on, great rents in the luminous envelope of the Sun, of what an immense capacity these enormous gulfs, those gigantic abysses, must be; our entire globe would appear in their depths no larger than a fragment of rock rolled into the crater of a volcano. With such dimensions as these, its spots must sometimes be visible to the naked eye.'[3]

Arago mentions spots seen equal to ten times the diameter of the Earth.[4] Dr. Young says that the largest spot yet recorded was observed in 1858. It

---

[1] See drawing of one observed by Professor Secchi, Guillemin's *Sun,* p. 221.
[2] Lockyer, pp. 226, 227. [3] Guillemin's *Sun,* p. 178.
[4] Arago, p. 409.

had a breadth of 143,000 miles, and covered about one thirty-sixth of the whole surface of the Sun.[1] Mr. Proctor also mentions the sizes of various spots, stating the largest to have been 143,500 miles.[2]

Williams, referring to the great spot observed in 1839, above mentioned by Guillemin, says: 'Worlds of the size of ours might be poured by hundreds into such a cavity, like peas in a basin.'[3]

2. The spots which advance from east to west sometimes '*preserve nearly the same form for many days together*,'[4] but as a rule they *are always changing their aspect*, are formed and closed with tremendous velocity; and ' it not unfrequently happens,' says Schellen, in his work on spectrum analysis, ' that the appearance of a group of spots is so entirely changed from one day to another that it can no longer be recognised in the new form it has assumed.'[5] When near to each other, says Arago, ' the nuclei appear to have a certain tendency to unite together. They usually enlarge till the moment when their union is effected.'[6] I have little doubt that this tendency is the result of gravitation between two or more comparatively solid bodies of vapour, which the nuclei must be. Sir John Herschel says:—

'When watched from day to day, or even from hour to hour, they appear to enlarge or contract, to change their forms, and at length to disappear altogether, or to break out anew in parts of the surface where none were before. In such cases of disappearance the central dark spot always contracts into a point, and vanishes before

---

[1] Young's *Sun*, p. 126.   [2] Proctor's *Sun*, p. 233.
[3] Williams, p. 112.   [4] Schellen, p. 276.
[5] Schellen, p. 275.   [6] Arago, p. 435.

the border.' 'Spots have been observed whose lineal diameter has been upwards of 45,000 miles, and even, if some records are to be trusted, of very much greater extent. That such a spot should close up in six weeks' time (for they seldom last much longer), its borders must approach at the rate of more than 1000 miles a day.'[1]

The appearances of a spot breaking up into smaller fractions are described by Francis Wollaston, in 1774, as 'similar to those which happened when, after having hurled a mass of ice upon the surface of a frozen pond, the different fragments into which it divides slide off in all sorts of directions,'[2]—a description which Guillemin says must not be taken too literally.[3] Dr. Long saw a spot not much less in diameter than our Earth break into two—the parts separating 'from one another with prodigious velocity.' The rate of separation or union, however, has been calculated. Guillemin says:— 'According to the recent observations of M. Chacornac, small spots are precipitated into larger ones with a velocity which sometimes attains to $599\frac{1}{2}$ yards per second,'[4]—an exactitude of measurement even to half a yard not a little surprising, although it in reality means a rapidity of 20 miles per minute, equal to 28,800 miles in a day, or thirty times the speed of a flying railway train. The same author mentions two cases of spots which M. Langier estimated receded from each other with a velocity of 121 yards per second, equal to about 5800 miles per day.

3. A *third* circumstance is, that the spots are only to be found in *two zones or belts near to the equator of the Sun*. They are never observed at the poles of the Sun,

---

[1] *Outlines*, § 386.   [2] Arago, p. 434.
[3] Guillemin's *Sun*, p. 191.   [4] *Ibid.* p. 191.

which, I may incidentally mention, are not thought to be much colder than the region of the Sun's equator. The equator itself is generally clear of spots, as may be noticed by examining Carrington's Plates of Observations.

4. But a much more singular circumstance is, that the spots are found to travel round the Sun, or accompany it in its rotation on its axis, at *different rates of speed*, dependent on their proximity to or distance from the Sun's equator.[1] The nearer they are to the equator they travel the more rapidly. This observed fact has explained the difference of observations as to the period of the Sun's rotation, which, having been computed by watching the period of revolution of the spots, has varied greatly.[2] It would also evidence a great want of coherence in the photosphere of the Sun, and probably, though Mr. Lockyer thinks not, in its whole body, supposing the Sun to be entirely gaseous, and no solid nucleus within.[3] The fact was discovered by Mr. Carrington, who spent seven years and a half in watching with painstaking minuteness[4] the appearances of the Sun; and the result of his labours on this point was, according to Mr. Lockyer, that he found the rotation of the Sun at the equator is

---

[1] Professor Newcomb, however, holds a different opinion. He says 'that the equatorial regions of the Sun perform their revolution in a shorter time than those parts nearer the poles, cannot be regarded as a scientific theory' (Newcomb, p. 251).

[2] Lockyer, p. 66.   [3] *Ibid.* p. 39.

[4] Mr. Carrington's labours are exhibited in a large quarto volume, published in 1863, *Observations of the Spots of the Sun*, November 9, 1853, to March 24, 1861, made at Redhill. The plates, 166 in number, embrace diagrams showing 99 rotations. Sometimes these diagrams are clear altogether of spots.

in 30·86 days, or nearly 31 days;[1] while at 50° N. or S. latitude (the highest points at which spots have been observed) the rotation takes place in 28·36 days, or in about 28 days, 8 hours.

5. Another peculiarity of the spots arises out of the influence exerted on the Sun by the *relative position*, the approach or retirement, *of the planets*, and by their conjunctions, and especially by the movements of the planets Mercury, Venus, and Jupiter.

'In order to obtain grounds for this conclusion,' says Mr. Lockyer,[2] 'the Kew observers have laboriously measured the area of all the Sun spots observed by Carrington from 1854 to 1860, and they find as the result of their inquiries that a spot has a tendency to break out at that portion of the Sun which is nearest to the planet Venus. As the Sun rotates, carrying the newly-born spot farther away from this planet, the spot grows larger, attaining its maximum at the point farthest from Venus, and decreasing again on its approaching this planet. We here speak of Venus, as it appears to be the most influential of all the planets in this respect. Jupiter appears also to have much influence, and more recently it has been shown that Mercury has an influence of the same nature, although more difficult to discuss on account of his rapid motion. Should, therefore, any two of these planets, or, still better, should all three be acting together at the same place upon the Sun, we may expect a very large amount of spots, which will attain their maximum at that portion of the Sun most remote from these planets.'[3]

---

[1] I am at a loss to reconcile this with Mr. Carrington's *Observations*, which throughout seem to take the rotation to be about 26 days. Dr. Young (*Sun*, p. 133) puts the period of rotation down at only 25 days.

[2] Lockyer, p. 81.

[3] Dr. Young considers it more probable that the periodicity of spots is in the Sun itself dependent upon no external causes, but upon the

It follows as a corollary that the more planets come in conjunction the greater will be the number of spots or disturbance of the Sun. Mr. Williams states the mean gravitation of Jupiter upon the Sun to be 13 times that of the Earth; of Venus, about $2\frac{1}{2}$ times; and of Saturn, a little above equal; while that of the other planets, Uranus and Neptune, is considerably less than that of the Earth.[1]

6. After the statement of these facts you will, no doubt, be prepared for another—that the *number of spots* during successive years seems to run a course, or what is termed *a cycle*, increasing gradually to a *maximum*, or largest number, and diminishing again to a *minimum*, or the smallest number, and so on. Looking to the regularity of the planetary motions (to which the fact must, I think, be mainly ascribed), we should naturally expect this result; and to the consideration of this question the attention of astronomers has been, and is still being directed. Mr. Lockyer mentions that Herr Schwabe of Dessau ' has now for about 40 years been engaged, without intermission, in registering the number of spots which appear on the Sun's surface,'[2] and the result of his labours has been to declare for a period or cycle of 10 years. In other words, that the same series of spots are formed and present themselves after a lapse of 10 years.[3] This conclusion, it is only right to say, has been disputed on the ground that it does not, as alleged, tally with recorded observations;

---

constitution of the photosphere, and the rate at which the Sun is losing heat (Young's *Sun*, p. 152).

[1] Williams' *Fuel of Sun*, p. 57.   [2] Lockyer, p. 79.

[3] A table of the number of days in each year without spots—from 1826 to 1869—is given in Proctor's *Sun*, p. 199.

and it cannot be positively affirmed that we are yet in a position to lay down authoritatively a law on the subject, or rather to fix the precise period of time, although it is commonly reckoned to be 10 or 11 years. As all the planets, and even the distant ones (which are the larger bodies), exert an influence more or less, and as their relative positions and conjunctions are intricately varying, it is probable, I think, that there may be a general law giving a 10 or 11 years' period due to the positions of Mercury, Venus, the Earth, and Jupiter; or, perhaps, according with Jupiter's period of revolution (which occupies 11 years 314 days), subject to displacement by the disturbances occasioned by the changing positions of the other planets, whose influence, however, individually is greatly smaller, though conjointly, as has been the case during the last 3 years, it becomes important. Indeed, it is considered that the 11 years' period is part of a larger cycle of 33 years, while some have stated another period of maximum spot development of about 56 years, corresponding to the epoch at which Jupiter and Saturn come into conjunction.[1] The fact, however, is a very weighty one, and, as I shall immediately show, we dwellers upon the Earth are deeply interested in it. In the meantime I must take up an equally interesting topic.

THE SOLAR PROMINENCES.

These are only visible during a total eclipse, and they are of irregular form and of lake red colour. They were at first supposed to belong to the atmosphere of the Moon; but that they are appearances in the solar atmosphere is now regarded as beyond question.

[1] Williams, pp. 55, 56.

The first recorded observation of these red flames was made at Berne in 1706. They were again noticed in 1715 and on subsequent occasions; but they were particularly watched during the eclipse of 1842.[1] M. Baily, one of the observers, thus describes them:—

'They had the appearance of mountains of a prodigious elevation; their colour was red, tinged with lilac or purple; perhaps the colour of the peach blossom would more nearly represent it. They somewhat resembled the snowy tops of the Alpine mountains when coloured by the rising or setting Sun.'

Another writer compared the appearance to 'a box of ebony garnished with rubies;' and another observed them 'change from white to red and from red to violet, and then back again through the reverse order.' The rapid changes they underwent was a feature of the observation of this eclipse. In 1851, they were again watched, and for the first time photographed, although photography does not appear capable of furnishing views of the Sun as distinct and clear as can be got by the eye itself.[2] One of the prominences then observed was so brilliant as to be visible to the naked eye. But it would occupy too much space to describe the appearances during this and subsequent eclipses. By that of 1860, when photography was better applied, it was established that there was a continuous envelope (the chromosphere) surrounding the Sun, and that the prominences were merely local heapings up of the envelope, and were accordingly attached to the Sun,—a fact previously conjectured, if not proved, by Mr. Grant in his history of physical astronomy.[3] These observations led to the discovery that it was possible to photograph

---

[1] Lockyer, p. 108.   [2] See Lockyer's *Star-gazing*, pp. 476, 477.
[3] Grant's *Physical Astronomy*, pp. 395–401.

the prominences while they were invisible to the eye, and even while the Sun is not under eclipse.

Mr. Lockyer in 1866 ingeniously suggested the possibility of observing the prominences when there is no eclipse of the Sun, and after working with a spectroscopic instrument of insufficient power, was, on 16th October 1868, put in possession by means of a Government grant of a better one.[1] Four days after he was enabled, he tells us, to examine the prominences while 'the Sun was shining in all the glory that an English autumn permits,' and with the result of discovering the existence of that for which he was looking, 'bright lines' in the spectroscopic image. An eclipse had happened in August of the same year, and had been watched in India by M. Janssen, the French astronomer. He also observed the bright lines, and being struck with the brilliancy of the red flames he was led to consider 'it might be possible to see them' (the spectroscopic image) 'without an eclipse.'[2] He tried it next day, and was rewarded by the sight, and for seventeen days afterwards with similar success. Although he had the good fortune to be in point of time earlier than Mr. Lockyer, who had first suggested the possibility of the examination, yet his communication to Paris was not received till Mr. Lockyer had published his independent discovery.[3] The scientific importances of these concurring observations was this, that they settled, at least in great measure, the problem of what do these flames or prominences consist.[4]

It had been previously thought that the prominences

---

[1] Lockyer, p. 127. [2] *Ibid.* p. 127.
[3] Mr. Proctor seems mistakingly inclined to give M. Janssen the greater credit (*Sun*, p. 280).
[4] Lockyer, p. 77, 78.

were not due to burning vapour, and that they might be particles of solid matter at a red glowing heat suspended in the atmosphere. The question being whether 'the particles in the photosphere itself might not be likened to a white hot poker, and those in the atmosphere to merely a red hot one.' The bright lines now discovered upset this idea, and showed that the prominences were truly burning gas. But I shall best convey a description of the observation in Mr. Lockyer's own words.

'Three beautifully coloured lines of light were visible. Two of them, corresponding to C and F in the solar spectrum, showed that the famous red flames were composed in part at least of incandescent hydrogen gas; that hydrogen gas was present in the atmosphere of the Sun in volumes beside which the size of the Earth is as nothing, welling up in what may be almost considered tongues of flame to a height of 70,000 and 80,000 miles, now running out into strange shapes and branches, now parting from the lower surface, and floating cloudlike in the higher regions. Besides these two lines which settled the question as to hydrogen, another line was observed near D which, strangely enough, had no dark line in the solar spectrum corresponding with it. I soon found that by sweeping the slit of the spectroscope along the Sun's edge and over the prominences it was quite easy to determine their outline, the length of the bright line visible giving the height of that part of the prominence on the slit at the time.'[1]

M. Janssen's observations were nearly similar, and he declares as a result of his investigations that the prominences 'are the seats of movements of which no terrestrial phenomena can give any idea, masses. of matter many hundred times larger than the Earth changing both place and form in the space of a few minutes.'[2]

[1] Lockyer, p. 125.   [2] *Ibid.* p. 128.

The rapidity of the changes and the height to which the prominences have reached, or been blown or driven, has been matter of repeated observation. For example, a prominence 40,000 miles high was subjected to such forces as in 10 minutes to leave scarcely a vestige of it—the changes being apparently greatest in the higher and rarer atmosphere.

Professor Respighi, an Italian astronomer, who has devoted himself with enormous labour to the observation of the solar prominences, states that he never found a prominence sensibly higher than 6′—that is, 150,000 miles; and he has furnished a table of the comparative heights, from which it appears that out of 1363 prominences, 1154 did not rise higher than from 25,000 to 50,000 miles, and only 5 exceeded 125,000 miles.[1]

Dr. Young says he has—

'Seen in all perhaps three or four which exceeded 150,000 miles; and Secchi has recorded one of 300,000 miles. On 7th October 1880, the writer observed one which attained the hitherto unprecedented height of over 13′ of arc, or 350,000 miles.'[2]

Most astronomers divide the prominences into two classes—eruptive and nebulous; but Respighi, from the extent of his observations, is inclined to subdivide much further. They are evidently due to a *welling up* of hydrogen and other vapours or gases from the hotter regions below, and have been ascribed by some to electricity, by others to heat, and by others to explosions.[3] Whatever be their cause, they are thus the counterparts of the spots which are due, as already stated, to the *downrush* of metallic vapours from the cooler heights above. Indeed, so closely connected is this action (which may, I presume, be somewhat likened to that

[1] Lockyer, p. 395.   [2] *Sun*, 202.   [3] Lockyer, p. 402.

which goes on in the ocean between the poles of the Earth and the equator) that—

'As a rule, when we get the bright line indication of an uprush we get an absorption line' (indicating downrush) 'by the side of it, often moved towards the red, which shows that we have relatively cooler hydrogen descending from above the disturbed part, but at times the index line in many cases disappears altogether,—that is, we have no longer relatively cooler hydrogen,—the whole of the superincumbent hydrogen has been heated to the same temperature as that of the newly ejected material, which is as hot, or at times hotter than the photosphere.'[1]

The prominences rise with immense rapidity, an ordinary rate being 40 miles per second; but often vastly more rapid. Conceive this if we can. Williams says:—

'It must be an enormous volley—millions of cubic miles of a furiously explosive mixture—consisting chiefly of the elements of water.' 'It must be a *continuous* explosion of such mighty force and magnitude that beggars the imagination in its efforts to picture its action and results. We know the deafening report which accompanies the explosion of a soap bubble when filled with these gases accurately mixed in the proportions to form water. What would be the crash if the cupola of St. Paul's were filled in like manner and exploded!' 'What if a hundred worlds all charged with the material of this horrid ruin were fired successively in one long bellowing train, combining their propulsive efforts like the contiguous grains of powder in a gun!'[2]

But not only do the prominences rise with such inconceivable fury and velocity, but they change their form

[1] Lockyer, p. 324.   [2] *Fuel of Sun*, p. 111.

in the most marvellous manner. One of the most extraordinary effects witnessed by man was one which happened on 7th September 1871, and has been described by Professor Young.[1]

'Just at noon the writer had been examining with the telespectroscope an enormous proturberance or hydrogen cloud on the eastern limb of the Sun,'—'a long, low, quiet-looking cloud not very dense or brilliant, nor in any way remarkable except for its size,' — 'floating above the chromosphere with its lower surface at a height of some 15,000 miles, but was connected to it, as is usually the case, by three or four vertical columns brighter and more active than the rest.' 'It was about 100,000 miles long by 54,000 high.' On viewing it again, in less than half an hour, he 'found the whole thing had been literally blown to shreds by some inconceivable uprush from beneath, and the air was filled with flying debris.'

When he first looked, some of them had already reached a height of nearly 100,000 miles, and while he watched them they rose with a motion almost perceptible to the eye, until in 10 minutes the uppermost were more than 200,000 miles above the solar surface.

As the filaments rose they gradually faded away like a dissolving cloud, and at 1.15 only a few filmy wisps, with some brighter streamers low down near the chromosphere, remained to mark the place.

What a wonderfully fascinating spectacle the Sun would present to us on some planet, were there one, near enough to enable us, without being burnt up, to witness such explosions, and perhaps to hear their terrific sound! How inexpressibly beautiful, but how awful would such appearances be! How overwhelmingly grand, but fear inspiring, would we feel those prodigiously

[1] Lockyer, p. 398.

colossal manifestations of the innate forces of nature exerted in play without an effort and unconscious of their tremendous power,—a power controllable only by the Creator from whom it was received! How trivial beside them would seem the most magnificent pyrotechnical displays of the creature man!

Professor Young, who says the velocity often exceeds 100 miles a second, and sometimes, though rarely, reaches 200 miles, calculated the uprush he described as just mentioned to be at the rate of 166 miles per second. But let us think for a moment what this terrific rush of 166 miles per second means. It is close upon 600,000 miles per hour. We have all seen and started back from an express train whizzing past us at the rate of 40 miles per hour, and know what that means. Yet supposing it 60 miles per hour, nearly the highest speed, the rapidity of the solar uprush of furious flame was ten thousand times greater. We cannot take it in, strain our imagination as we may. The downrush which creates a spot is probably not so rapid, but it must bear some proportion to the uprush. We have seen that the spots form and close at the rate of 1000 miles a day; but the Sun is swept by cyclones with which we have nothing on earth to compare.[1] For the solar wind is calculated to speed at the rate of 120 miles per second, while the greatest hurricane on earth has only blown at the rate of 100 to 120 miles per hour.[2]

And it is not the mere speed of rush upwards or downwards or along the surface of the Sun which is to be regarded; nor is it a rush confined to spots or

[1] Lockyer, pp. 236, 401.
[2] Buchan's *Meteorology*, § 410; Hartwig's *Aerial World*, p. 111.

to *faculæ*, for probably every willow leaf is hissing with the light and heat which wells up through it to the surface, and every interstice, every pore or space between the leaves is whistling with the downrush of hot metallic vapour seeking to regain its home in the regions below. Remember, too, what a frightful element it is which is being tossed up and down, back and fore—the element of fire—of heat and light beyond our faintest conception. What a tempestuous burning ocean must be for ever boiling and seething and tossing, blown and hurled with irresistible fury into waves perhaps higher than our loftiest mountains, sublimely grand amidst an uproar which would appal the shuddering heart of the stoutest demon of the fiery lake.

M. Flammarion in a popular work, alluding to the solar spots, says :—

'We may be led to the belief that these are insignificant movements carried on on the Sun's surface, and of small extent. It is not so. They are daily and very important phenomena. Some of them have been known to measure 80,000 miles, that is to say, they are ten times larger than the Earth. Our globe falling into most of them would be lost as in a well. Besides being of this size, they are also the seat of various actions and prodigious phenomena. They are not formed suddenly as a whole, but increase to the limit they attain, and afterwards diminish. Some only last a few weeks, others months. Now the movements with which they are animated, either for their increase or diminution, or in their internal action, are sometimes of unheard of rapidity. Lately astronomers have followed a dazzling meteor passing through a group of spots with a velocity of 2000 French leagues per minute. In other parts they have watched circular whirlwinds dragging into their commotion large spots like the Earth, and swallowing them up in abysses with fearful velocity. Sometimes are seen the crests of stormy waves extending over parts of the penumbra, and rising on the white surface of the Sun as a still whiter or brighter substance, doubtless projected in their ebullition by interior forces. There have, besides, been seen immense bridges of fiery substances cast suddenly over a black spot crossing it from one end to the other, like an arch of luminous striæ, which sometimes is dissipated, and falls down into the abysses of lower whirlpools. This body, which each day pours out over our heads such a pure and calm light, is the seat of powerful actions and prodigious movements of which our tempests, hurricanes, and water-spouts give us but a slight idea; for these gigantic disturbances are not performed, as here, in an atmosphere of a few leagues' thickness and over a few leagues' area, but in proportions as vast as its atmosphere, which rises thousands of leagues above its surface, and as its volume, which exceeds 1,450,000 times that of our globe.'[1]

Now do not suppose that these disturbances in the Sun are immaterial to us dwellers upon Earth. For every movement in the Sun is fraught with influence upon the Earth. The greater the action the greater the heat evolved. The goodness or the badness of our seasons is dependent upon the extent of the Sun's convulsion. Mr. Proctor, in one of his many books (*Other Worlds than ours*, published in 1870), after stating certain of the effects of what he terms the magnetic action in the Sun upon the Earth, and referring particularly to a magnetic storm in the Sun observed on 1st September 1859, which sharply disturbed the self-registering magnetic instruments of the Kew Observatory, interrupted telegraphic communication, and in some cases set the telegraphic offices on fire, while ' auroras appeared both in the northern and southern

[1] Flammarion, *Marvels of the Heavens*, p. 145.

hemispheres during the night which followed, and the whole frame of the Earth seemed to thrill responsively to the disturbance,' says :—

'No disturbance can affect the solar photosphere without affecting our Earth to a greater or less degree. But if our Earth, then also the other planets, Mercury and Venus, so much nearer the Sun than we are, surely respond even more swiftly and more distinctly to the solar magnetic influences. But beyond our Earth, and beyond the orbit of moonless Mars,[1] the magnetic impulses speed with the velocity of light; the vast globe of Jupiter is thrilled from pole to pole as the magnetic wave rolls in upon it; then Saturn feels the shock, and then the vast distances, beyond which lie Uranus and Neptune, are swept by the ever lessening yet ever widening disturbance wave.'[2]

The influence[3] thus poetically but truly ascribed by Mr. Proctor to the Sun was an opinion broached so early as 1614 by Baptiste Baliarios, an Italian mathematician, and was revived by Sir William Herschel, but for want of meteorological observations he and others were obliged to take the prices of wheat as the basis (a very rough one) of comparison and deduction, and it was not surprising that very opposite conclusions were formed by eminent philosophers. While Herschel himself did not escape ridicule for his speculations, a writer in the *Edinburgh Review* thus delivering himself :—

'To the speculations of the doctor on the nature of the Sun we have many similar objections, but they are all eclipsed by the grand absurdity which he has there

---

[1] Mars is now discovered to have two Moons, but of small size.

[2] Proctor, p. 34.

[3] See subject discussed in Herschel's *Popular Lectures on Scientific Subjects*, p. 79 (1876), and Skertchly's *Universe*, p. 212.

committed in his hasty and erroneous theory concerning the influence of the solar spots on the price of grain. Since the publication of *Gulliver's Voyage to Laputa* nothing so ridiculous has ever been offered to the world.'

It is only right, however, to say that Mr. Carrington does not seem to attach importance to calculations based on price of grain, because ' social and political causes affect prices to an extent sufficient to destroy their value for the purpose for which he (Herschel) selected them.'[1]

But the subject has of late years been more fully investigated, and the results form a new chapter in meteorology not to be found even in recent books on this branch of science.

' Mr. Meldrum, a distinguished meteorologist, who lives in the torrid zone,' says Mr. Lockyer,[2] ' tells us that the whole question of cyclones is a question of solar activity, and that if we write down in one column the number of cyclones in any given year, and in another column the number of Sun spots in any given year, that there will be a strict relation between them—many Sun spots, many hurricances; few Sun spots, few hurricanes. Only this morning I have received a letter from Dr. Stewart, who tells me that Mr. Meldrum has since found that what is true of the storms which devastate the Indian Ocean is true of the storms which devastate the West Indies; and on referring to the storms of the Indian Ocean, Mr. Meldrum points out that at those years where we have been quietly mapping the Sun spot maxima the harbours were filled with wrecks, vessels coming in disabled from every part of the great Indian Ocean. Now that is surely something worth considering, because if we can manage to get at these things, to associate them in some way with solar activity so that

---

[1] *Observations*, p. 248.   [2] Lockyer, p. 423.

there can be no mistake about it, the power of prediction—that power which would be the most useful one in meteorology, if we could only get at it—would be within our grasp.'[1]

Observation has likewise been directed to the correspondence of particular solar paroxysms with particular storms upon Earth,[2] and it has been stated that the one follows the other after the lapse of above *two hours of time.* The rainfall has similarly been found, according to tabulated statements, to be obedient to the same influences.[3] But there is a difference of opinion as to whether the terrestrial phenomena are most violent when the spots are at their *maxima* or at their *minima.* Professor Balfour Stewart holds that there is greatest solar activity when the spots are greatest (which, so far as I can pretend to judge, really appears to coincide with observation), and that this activity is greatest when the planets are in conjunction, or when they are nearest the Sun, is most likely. Indeed, a Dr. Knapp of Chicago has published, in 1879, a pamphlet (reprinted in Edinburgh) called *Coming Disasters on the Earth, 1881 to 1885, resulting from the Perihelia of the Planets.*[4] For (taking Dr. Knapp's dates) Jupiter attained its perihelion in October 1880; Mars, in May 1881. Neptune was due in end of 1881; Uranus will reach its perihelion in spring of 1882; and Saturn in autumn of 1885. Besides, in 1881 and 1882 Jupiter, Saturn, and Neptune have been, and are, and will be almost in conjunction, so that their combined

---

[1] The severe weather experienced in December 1882 followed the appearance of a great solar spot, which possibly may have caused the storm of snow and cold.
[2] Lockyer, pp. 639, 640.   [3] *Ibid.* p. 426.
[4] Edinburgh: Robert Somerville, 1881.

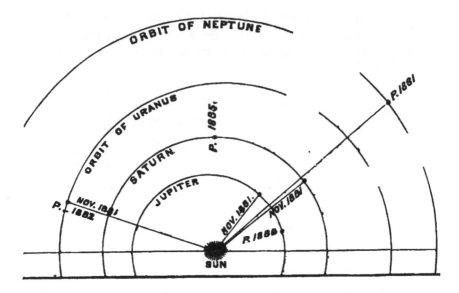

influence on the Sun is great.[1] Such an event 'as four of the largest planets being in perihelia within so brief a space of time as five years' is a combination which is said not to have happened for 2000 years, and serious results—storms, floods, earthquakes, and extremes of heat and cold, producing famines and plagues—might, the pamphlet said, be expected; and in proof Dr. Knapp details a correspondence between such perihelia in former years and famines, pestilences, and other disasters on Earth, although he does not seem to trace them up to observed disturbances in the Sun. The severe and changeable weather we have, unfortunately, experienced during the last two or three years, embracing days of 'the greatest cold that ever was known in the British Isles,'[2] certainly seems to support the

[1] The above diagram has been adapted from that of Dr. Knapp, taking his dates of the perihelia, and drawing lines to show the positions at November 1881. Assuming the dates to be correct, it will be seen how closely the perihelia approximate.

[2] Mr. Buchan's statement at Edinburgh Botanical Society, November 1881.

conclusion of its direct connection with solar perturbation arising out of the position of the planets. Indeed, what could be more convincing than the storms which have visited us during the last few months (the latest only last night), when Neptune has been attaining his perihelion. We can well imagine that this distant planet, conscious of his wanderings from his great parent, and agitated by his near approach,—nearer than he has been for 165 years, the period of his revolution,—and distracted by the tugs of his big brothers, Jupiter and Saturn, stoiters and staggers in his course, and, doubtless, is drawn many millions of miles nearer to us than he would wish, or than would otherwise be. While all the three, Jupiter, Saturn, and Neptune, are having a long pull, a strong pull, and a pull altogether upon their susceptible father, the Sun, who, fancying the time has come in which to gather his erring offspring to himself in a devouring embrace, is, with intenser light flashing from his eyes, opening his soft but treacherous heart to receive the guilty truants, and, excited by the animating thought into a tumult of boisterous emotion, sends a tremor of terror thrilling to us and to all the trembling worlds of his system. It is confirmatory of this view to know that the Sun has been lately filled with spots. Some days ago Mr. William Southern, F.R.A.S., writing from Leamington to the *Daily Post* on 23rd November 1881, says:—

'It will doubtless interest many of your readers possessing telescopes of even small aperture to know that just now the Sun is full of interest on account of the remarkable number of spots visible. On Monday last I counted twenty. A group very much resembling the configuration of the constellation Ursa Major—Great Bear—occupied a position a little north of the solar equator, and about midway between the eastern and western limbs.'

Mr. Southern refers in his letter to the eleven years' cycle, but makes no allusion to the more important fact of the planetary conjunction and perihelia. The whole subject, however, of the connection between solar spots and terrestrial weather has been for some time matter of investigation and discussion, although apparently little attention seems to have been directed to the planetary influences now affecting us. I can only refer to it thus generally in the expectation that ere long some one qualified to deal with the matter will give us full enlightenment. Meantime I may just observe that, looking to the position of the major planets, our farmers and our sailors may for some time to come—possibly even for a few years to come—continue to experience unfavourable and perhaps stormy weather. But when this severe and trying period shall have passed, it is satisfactory to think that probably it may not again occur in the world's history till the next similar conjunction, perhaps after another long 2000 years shall have run their eventful course.[1]

It is not a little curious to find that the ancient Chaldeans 'had an astro-meteorological period consisting of twelve solar years in which they conceived the phenomena of weather to recur.'[2] To a somewhat similar opinion the moderns are now inclining, but doubtless on better scientific grounds. As already stated, the cycle of Sun spots has been supposed to be a period of ten or eleven years, and may possibly be,

[1] This lecture was delivered on 7th December 1881, and severe storms and floods afterwards happened.
[2] See G. C. Lewis' *Astronomy of Ancients*, p. 311.

in accordance with Jupiter's revolution, more nearly the Chaldean period of twelve years. It is therefore most important for us to obtain a true knowledge of the law of succession of these solar disturbances, because we should then obtain a clue to so far unveiling the mysteries of nature and obtaining a foreknowledge of those terrestrial changes which may enable us to foresee, to some extent, like Joseph, divinely inspired, a seven years' plenty or a seven years' famine. Imagination has indeed been already sufficiently lively in attributing to solar activity such consequences as the American Secession War and our periodic commercial crises. But more reliable and practical deductions are being made. For, as we all know, our mariners have greatly benefited by the forecasts of storms which our American brethren so kindly and so frequently telegraph to Europe. When our knowledge shall have become more complete and exact, meteorology may have attained a higher place, and will confer greater benefits upon man.

Looking to all that has been now stated regarding the Sun, yet without entering at present on the question, it would require some confidence to affirm that the Sun is the abode of life.

But let us suppose it possible, and let us imagine, with a free indulgence in poetical licence, some being akin to man carried to the Sun's surface.

One of the first effects he would experience would be to find every shred of clothing burnt off his body long ere he entered the Sun's atmosphere. Well, that would not signify much. There would be little fear of any cold shiver supervening. Matters far more alarming would at once confound him. Every drop of blood would be instantaneously evaporated. Dried, parched, scorched to a living cinder, he would most resemble a bit of unburnable asbestos,—with, however, we must suppose, in order that he might comprehend his position, all his senses intact, and, among others, those of feeling, hearing, and also sight, which the glare of the Sun, even encountered 90 millions of miles off, takes away. Suppose this remarkable, imperishable, glowing, red being,—this desiccated, burning, curled up cinder, suffering torture inconceivable, steps upon the photosphere of the Sun, and with a yell of agony—if a yell could be uttered, and, if uttered, could be heard—sits down upon a hot willow leaf larger than Great Britain. In torment beyond conception he looks around. There is not a living being of any sort with whom to fraternize in woe. A terrible solitude of devouring anguish environs the place. He looks up through a blaze of light. A comet on lightning wings is rushing and flashing past. Meteors in millions are gleaming and bursting and falling in fiery showers fast and close and furious. There is not a particle of blue sky on which to rest his parched and blinded eyes. He looks around, and there is not a green tree, or even a green blade of grass, to refresh his frenzied gaze. There is not a drop of cold water to allay the feverish rage of his unquenchable thirst. There is not a cool cave or grotto into which to retreat from the dazzling glare, or from the overpowering conflagration into which every element of the Sun is thrown. There are mountains indeed, and mountains of gigantic height, but they are masses of incandescent metal,—rather are they phantom mountains of thick fervent vapour, heated to a degree with which we have nothing on Earth to compare. There are rivers indeed, or what seem to be so, but they are to appearance torrents of liquid metallic fire, not little, sputtering, crackling streams, but pouring heavily ocean wide in steaming, heaving, burning, tossing, angry, fearful waves, rolling over precipitous heights to awful hollows, and thundering with the roar of ten thousand hurling avalanches. With the horrid din of which, and of the loud tumultuous hissing of violent currents of wrathful gases tearing from their beds, bursting, blazing, and twisting as they mingle to fight in fierce raging conflict, and darting in colossal tongues of red flame 100,000 miles up with almost the swiftness of light, the bewildered ear is rent. And now a frightful, fiery, tempestuous tornado sweeps across his leaf with appalling vehemence. He is turned over and over, vainly clutching and clinging to ghostly shapes, and rolls, bruised and crushed, he knows not where. Everything around him he fancies grating and cracking, tearing and separating, and he feels his leaf is blown like a feather, and, with the speed of an express, is flung forward 10,000 miles, and tossed now up, now down, amid a crowd of others shaking and crushing for room, till at last he finds it suspended over a vast yawning, horrible gulf. In consternation, clinging to some burning barrier, he looks down the dreadful aperture, and sees a seething mass of impetuous scorching clouds, boiling and swirling and pouring to the bottom, and down, down, far down into the depths below, he fancies he descries a glowing metallic looking mass, which he imagines to be the great body of the Sun. Recoiling in fright, his leaf trembles under him, and, terror-struck, he looks up and finds he is again in motion; for a hurricane overwhelms the closing chasm, and he is being driven and sent spinning forward and forward, or rather backward and backward, to where he was, and all round about other gigantic glowing leaves like his own are impelled and wheeled as by some prodigious whirlwind; and amidst the noise of a Titanic conflict, clashing and crashing, they rush in wildest frenzy to embrace. Stunned by the awful shock, our living cinder is suddenly shot high into the air, and escapes from his terrific ordeal unconscious.

Such, could it be witnessed and experienced, might be the scene upon the Sun. Fortunately, we have nothing like it upon Earth. But to all this distant turmoil—which is death to all life upon the Sun itself—we owe the condition of our little planet, and its beautiful adaptation to receive organic life, and among and above all the being Man, to whom has been accorded not merely organic life, but the gift of intelligence and reason, and that which dignifies him beyond every other of the creatures upon Earth, a soul destined to live on and for ever on, when Sun and Moon and land and sea shall be no more.

PART II.

ARE THE HEAVENLY BODIES HABITABLE?

OR,

'THE PLURALITY OF WORLDS' CONSIDERED.

ARE THE HEAVENLY BODIES HABITABLE?

OR,

'THE PLURALITY OF WORLDS' CONSIDERED.

WHEN, on a clear night, we walk out into the open field, and, lifting our eyes to the azure vault, behold a thousand flickering stars scintillating above us, and know that, excepting a very few which are planets of our system, these are all vastly larger than the great globe on which we dwell; and when we reflect that for every shining orb which thus indicates its existence the telescope reveals to us millions more invisible to the unassisted eye; and that beyond the reach of all present, and perhaps of all future, telescopic power, the Universe is filled with other and like orbs, stretching in clusters and vast galaxies in every direction around, to the limitless verge of infinity; and when we conjoin with this grand thought the consciousness that all about on this Earth we see life prevailing in every conceivable shape,—the tendency is not unnatural mentally to conjure up the conception of numerous and diversified animal and vegetable forms in each of these myriad stars, and, in dreamy reverie, to think that upon each sparkling ball there are beings, active and intelligent, peopling and enlivening its every zone. Nay, while the distances, the magnitudes, the velocities, the periods of the heavenly bodies, or some of them, are ascertained by scientific means, with mathematical nicety, to a mile, to a ton, to a fraction of a second of time, but with results so vast that, knowing it all, our imagination refuses to realize figures so precise but so immense,—yet where by no application of the exact sciences, by no disposition of x and y, can there be demonstration of life, or even of a single unit of being, in any one of those lustrous spheres, that same imagination, no longer inert, seizes the bit, becomes ungovernable, and wildly rushes off with us, and not only endues each globe with life, but pictures to our glowing vision populations of men, of like forms and like passions with ourselves, engaged in all the pursuits of earth, and passing through all the phases of existence, from infancy to age, from ignorance to erudition; exhibiting here the old, old tale of true and tender and passionate love; toiling there, amid the din and bustle and anxieties of the overgrown city; or busily engaged from day to day in the hum and rattle of the noisy mill; or speeding, with rich and varied cargo, on the broad ocean, impelled by towering mast and swelling sail, or by the clanking engine and revolving wheel; or earnestly and skilfully peering at the wonders of the Heavens, aided by every mechanical power known to ourselves, and estimating by instruments of observation, of rare delicacy, their form, their distance, and their nature; or deigning in the secret chamber to raise to the ever-listening ear of the Most High the penitential prayer; or in the hallowed temple, in assembled throng of thousand voices, lifting in exalted strain the melodious measured hymn of grateful and adoring praise.

The thought is captivating; and here, if anywhere, amid all the marvels of astronomy, romance may fairly have its mission, and can afford to revel with exuberant delight. But our business is now, with calm consideration, to inquire on proper grounds whether imagination thus stirred has been rightly or wrongly directed; and, taking into view all the facts which philosophical investigation has brought to light, to form a sound and rational conclusion.

Yet ere we proceed to do so, it will be advisable, as briefly as possible, to glance at the views which from time to time have been entertained, and at the controversy which has been waged on this large, most interesting, and, as I doubt not it will be found, this most instructive subject. We shall then be in a position more thoroughly to comprehend and more easily to examine the various reasons which have in times past been assigned in favour of plurality, and from such examination to pass to consider in detail all the circumstances which bear upon the question, and especially those which influence determination regarding the various members of the solar system.

JOSEPH JEROME LALANDE, director of the Observatory of Paris, and the author of a treatise in three quarto volumes (which for a century was the standard work) on astronomy, thus expounds the conceptions of the ancients regarding our subject, and the history of opinion to the time of Fontenelle :—

'I have remarked, in the twentieth book of my Astronomy, that in every period of time it has been believed that the planets were inhabited, on account of their resemblance to the Earth. The idea of the plurality of worlds is expressed in the Orphics, those ancient Grecian poems attributed to Orpheus (*Plut. de Placitis Philosoph.* l. 2, cap. 13). Proclus has preserved some verses, in which we find that the writer of the Orphics places mountains, men, and cities in the Moon. The Pythagoreans—such as Philolaüs, Hicetas, Heraclides—taught that the stars were all worlds. Several ancient philosophers even admitted an infinity of worlds beyond the reach of our sight. Epicurus, Lucretius, and all the Epicureans were of the same opinion; and Metrodorus thought it as absurd to imagine but one world in the immensity of space as to say that only one ear of corn could grow in a great extent of country. Zeno of Eleusis, Anaximenes, Anaximander, Leucippus, Democritus, asserted the same thing. In short, there were some philosophers who, although they did not consider the rest of the planets inhabited, placed inhabitants in the Moon; such were Anaxagoras, Xenophanes, Lucian, Plutarch (*Di Oraculor. defectu. De facie in orbe Lunæ*), Eusebius, Stobius. We may see a long list of the ancients who have treated of the subject in Fabricius (*Biblio. Græcæ*, t. 1, cap. 20); and in the *Memoire de Bonamy* (*Acad. des Inscriptions*, tom. ix.) Hevellus appeared as firmly persuaded of this opinion in 1647, when he talked of the difference between the inhabitants of the two hemispheres of the Moon: he calls them *Selenitæ*, and examines at length all the phenomena observed on their planet after the example of Kepler (*Astron. Lunaris*). It was maintained at Oxford, in certain themes which are mentioned in the *News of the Republic of Letters*, June 1764, that the system of Pythagoras on the inhabitants of the Moon was well founded: two years afterwards Fontenelle discussed the subject in his agreeable work. There are further details of the different astronomical opinions at the end of Gregory's book. For the objections we may refer to Riccioli (*Almagestum*, tom. i. pp. 188, 204). In 1686, the *Plurality of Worlds* was adorned by Fontenelle with all the beauties of which a philosophical work was susceptible.'[1]

Historically, these notices are interesting; and the subject will be found more fully treated in Flammarion's *Mondes Habités*;[2] but, as adding to the real consideration of the subject, they are of no value. To the ancients, lacking the telescope, the Heavens were in reality a sealed book; and as they had no means of knowing what the stars are, and in writing of them only originated conceptions, often strange and fantastic, which they had no means of verifying, their speculations regarding the habitability were merely the dreams of philosophers who

---

[1] 'Critical Account of the Life and Writings of Fontenelle,' prefixed to translation of the *Plurality of Worlds*. This does not appear in all editions.

[2] Book I. ch. i. 24th ed. (1876), pp. 13–31.

knew nothing. Nor, indeed, were the earlier modern astronomers much better placed. They did know something more of the heavenly bodies, and formed more accurate suppositions regarding both their nature and their motions; but the science which they practised was in its nonage, and could afford no proper data from which to draw legitimate conclusions. Indeed, the 'little learning' they had was then, according to the saying,—though in a different sense from that in which it is commonly used,—'a dangerous thing.' For Ciampoli, in 1615, wrote to Galileo:—

'Put a great reserve on what you say; for when you establish a certain resemblance between the terrestrial globe and the lunar globe, another person immediately exaggerates it, and says you suppose *that there are men inhabiting the Moon;* and this other person soon begins to inquire how they can possibly have descended from Adam, or come out of the ark of Noah, with no end of other extravagances of which you have never dreamt.'[1]

From which it would seem, that to suggest the habitability of other planets was then likely to excite ridicule, while not improbably it might be found to be a perilous heresy.

During this infancy of astronomical science, various writers appear to have touched upon our subject, or to have expressed an opinion.[2] Nor had knowledge greatly increased when Fontenelle published his *Conversations on the Plurality of Worlds*, a book which has ever since identified his name with the question. Whatever may have been the merits or demerits of the work, it was for many long years read, credited, and honoured. It is

[1] Phipson's *Mysteries*, p. 266.
[2] See Flammarion, pp. 32-38.

indeed somewhat amusing to notice the attention it obtained, the panegyrics with which it was the fashion to commend it. Lalande says: 'Whenever I have entered into a conversation with any sensible woman on astronomy, I have always found that she had read Fontenelle's *Plurality of Worlds,* and that his book had excited her curiosity on the subject. As it has been so much read already, it must continue to engage attention.' Then mentioning the numerous editions through which it had passed in various countries, he quotes a sentence from Voltaire, declaring that he considered it one of the best books that ever was written. In a biographical notice of the author, prefixed to a French edition of Fontenelle's works, it is also stated that it had a prodigious success; and as the ladies adored Fontenelle, whose bon-mots and smart repartees enlivened the salons of Paris, one is not surprised to learn from it that 'les femmes surtout le lurent avec impressement et c'est encore (1825) celui de tous les ouvrages de Fontenelle qui est le plus généralement connu.'[1] Sir David Brewster, writing long after, in referring to 'this singular work,' says it 'excited a high degree of interest, both from the nature of the subject, and the vivacity and humour with which it was written. It was read with unexampled avidity, and was speedily circulated through every part of Europe. Wherever it was read it was admired.'[2]

The plan adopted by Fontenelle, as expressed in his preface, was to 'discourse of philosophy, but not directly in a philosophical manner, and to raise it to such a pitch that it shall not be too dry and insipid a subject

[1] *Œuvres de Fontenelle,* Paris, 1825. *Notice Historique sur Fontenelle,* vol. i. p. 28.
[2] *More Worlds than One,* p. 2.

to please gentlemen, nor too mean and trifling to entertain scholars.' In other words, it was intended to be the beau-ideal of popular philosophical treatment of the age in which it appeared.

On perusing it, however, one cannot help feeling surprised, not merely at the feeble shallowness of its reasonings, but at the marvellous scantiness of its astronomical facts. The few elementary points, inaccurately stated as historical truths, and enclosed within a cloud of a long exploded system of vortices projected by Descartes, who was then a living philosopher, are chopped up into spoon meat, and diluted by all the verbosities and digressions, the compliments and the badinage, the doubts, and the resolution of doubts, of concocted conversation; while into this diluted mixture a few indeterminate crumbs of argument are from time to time dropped, and lie soaking in empty inconclusiveness, till indeed it is difficult to extract precisely even the author's real views upon his great theme.

Apparently, however, his opinions were these. Holding the Sun to be a place not to be inhabited, he says the Moon is inhabited, because she is like the Earth; but, being also unlike, her inhabitants are of a different order; the other planets (for Moons in those days were called planets) are inhabited, because they are like the Moon. The fixed stars are like the Sun, and therefore not inhabited; but they have each a planetary system, the members of which must be inhabited; but comets, being reckoned as planets coming from other systems, are likewise inhabited. Very curious speculations are indulged in with regard to the Milky Way, which is supposed to be nothing but 'an infinity of small stars not to be seen by our eyes, because they are so very little, and they are sown so thick one by another that they seem to be one continued whiteness,'[1] and so near, 'that the people in one world may talk and shake hands with those of another; at least, I believe,' the author says, 'the birds of one world may easily fly into the other, and that pigeons may be trained up to carry letters as they do in the Levant!'[2] With all this superficiality of statement and reasoning, it is not surprising that the existence of inhabitants is a point assumed; leaving the Sage and his Marchioness free to discuss with all vivacity the wonderful character and qualities of those conjectural tenants of the various hot and cold celestial orbs or dwelling-places, to which, by the grace of Fontenelle, these tenants, willingly or otherwise, have been assigned a habitation.

But a still stranger, though generally less known work was shortly after written by the Dutch philosopher Christian Huygens, discoverer of the ring and one of the satellites of Saturn, and the inventor of, among other things, the pendulum in its application to clocks. For at the age of 65, Huygens spent his last days in writing the *Cosmothereos,* or, as it is called in the English translation, *The Celestial Worlds discovered; or, Conjectures concerning the Inhabitants, Plants, and Productions of the Worlds in the Planets;* and left instructions for its publication, which took place in 1698, after his death.[3]

Regarding this book, Sir David Brewster,[4] a keen advocate, as we shall see, for the plurality, wrote: 'The

---

[1] P. 133.   [2] P. 134.

[3] Fontenelle's work is not now to be found in every private library, but Huygens' is still more scarce. I could only see it in the library of the University of Edinburgh.

[4] *More Worlds than One,* p. 4.

Cosmothereos is a work essentially different from that of Fontenelle. It is didactic and dispassionate, deducing by analogical reasoning a variety of views respecting the plants and animals in the planets, and the general nature and condition of their inhabitants.' The work which Humboldt, taking a juster view, describes as consisting 'of the dreams and fancies of a great man,'[1] is of so singular a character, that I must, as briefly as possible, exhibit its contents and nature.

After referring to the affirmations of the ancients and of later authors, that the planets, and even the stars, are furnished with inhabitants, and saying that 'the ingenious French author of the *Dialogues about the Plurality of Worlds* had not carried the business any farther,' he imparts to the reader that there was room for his own conjectures; and, making certain explanations regarding the bodies of the solar system, he proceeds:—[2]

'Having thus explained the two schemes, there's nobody, I suppose, but sees that in the first the Earth is made of the same sort with the rest of the planets. For the very position of the circles shows it. And that the other planets are round like it, and like it receive all the light they have from the Sun, there's no room (since the discoveries made by telescopes) to doubt. Another thing they are like it in is, that they are moved round their own axis; for since 'tis certain that *Jupiter* and *Saturn* are, who can doubt it of the others? Again, as the Earth has its Moon moving round it, so *Jupiter* and *Saturn* have theirs. Now, since in so many things they thus agree, what can be more probable than that in others they agree too; and that the other planets are as beautiful and as well stocked with inhabitants as the Earth? or what shadow of reason can there be why they should not?

'If any one should be at the dissection of a dog, and

[1] Humboldt's *Cosmos*, iii. p. 22.   [2] Huygens' *Cosmothereos*, p. 17.

be there shown the Entrails, the Heart, Stomach, Liver, Lungs, and Guts, all the Veins, Arteries, and Nerves, could such a man reasonably doubt whether there were the same Contexture and Variety of Parts in a Bullock, Hog, or any other Beast, tho' he had never chanc'd to see the like opening of them? I don't believe he would. Or, were we thoroughly satisfied on the nature of one of the Moons round *Jupiter*, should not we straight conclude the same of the rest of them? So if we could be assured but in one Comet, what it was that is the cause of that strange appearance, should we not make that a standard to judg of all others by? 'Tis therefore an argument of no small weight that is fetched from Relation and Likeness; and to reason from what we see and are sure of to what we cannot, is no false Logick. This must be our method in this Treatise, wherein, from the nature and circumstances of that Planet which we see before our eyes, we may guess at those that are farther distant from us.'

He then concludes the probability of the planets being solid like the earth, and not wanting in gravity.

'But now,' he goes on to say (p. 19), 'to carry the search farther, let us see by what steps we must rise to the attaining some knowledge in the more private secrets concerning the State and Furniture of these new Earths. And first, how likely is it that they may be stocked with plants and animals as well as we? I suppose nobody will deny but there's somewhat more of Contrivance, somewhat more of Miracle, in the production and growth of plants and animals than in lifeless heaps of inanimate Bodies, be they never so much larger, as Mountains, Rocks, or Seas are. For the finger of God and the wisdom of Divine Providence is in them much more clearly manifested than in the other. One of the Democritus's or Cartes' scholars may venture perhaps to give some tolerable Explication of the appearances in Heaven and Earth, allow him but his atoms and motion; but when he comes to Plants and Animals, he'll find himself nonplus'd, and give you no likely account of their Production. For everything in them is so exactly adapted to some design, every part of them so fitted to its proper use, that they manifest an infinite Wisdom, and exquisite knowledge in the laws of Nature and Geometry, as to omit those Wonders in Generation we shall by and by show, and make it an absurdity even to think of their being thus haply jumbled together by a chance motion of, I don't know what, little Particles. Now, should we allow the Planets nothing but vast Deserts, lifeless and inanimate Stocks and Stones, and deprive them of all those Creatures that more plainly speak their Divine Architect, we should sink them below the Earth in Beauty and Dignity—a thing that no reason will permit, as I said before.'

Now follows a paragraph characteristic of the whole book; for having built up an argument on conjecture, he forthwith assumes a conclusion to be *proved*, and thence goes on, upon like suppositions, to infer and build up more:—

'Well, then, *now we have gain'd the Point for them*, and the Planets may be allow'd some Bodys capable of moving themselves not at all inferior to ours (for why should they?), and these are animals. Now, for fear of starving these poor creatures, we must have plants, you know. *And so the other point is gain'd.* And as for their Growth and Nourishment, 'tis no doubt the same with ours, seeing they have the same Sun to warm and enliven them as ours have.'

After qualifying this by stating that the plants and animals of the other planets[1] 'have indeed some difference in their *shape*, to distinguish the plants and animals of these countries from ours,' the author concludes the planets must each have water of a kind adapted to its individual heat, and *inter alia*, that the plants and animals must have modes of propagation

[1] Huygens' *Cosmothereos*, p. 23.

similar to those of Earth, and assuming[1] 'all this furniture and beauty the Planets are stock'd with,' says:—

'The Planets seem to have been made in vain, without any design or end, unless there were some in them that might at the same time enjoy the Fruits, and adore the Wise Creator of them.'

And then he proceeds on this, as a concluded basis, to detail that in all things these supposed rational creatures must resemble man. One among the many curious conclusions he forms is that the inhabitants must be astronomers, and being astronomers, must have all the implements and practise all the arts necessary to cultivating it, only, singularly enough, he 'dare not' allow them telescopes—[2]

'For fear people should be so disturbed at the ridiculous extravagancy of such an opinion as to take the measure of my other conjectures by it, and hiss them all off upon the account of this alone.'

Having thus to his own satisfaction 'gain'd our point, and 'tis probable that they are as skilful Astronomers as we can pretend to be,' he deduces they must have hands, feet, upright carriage, senses, and faces like ourselves.[3] In short, he 'cannot, without horror and impatience, suffer any other figure for the habitation of a reasonable soul' 'than our own.' In other words, the inhabitants must be men similar to ourselves, enjoying the pleasures of society, dwelling in houses, having ships, 'with sails, anchors, ropes, pullies, and rudders;'[4] 'and perhaps they may not be without the use of the compass too.' Geometry they must have, being a science 'of such singu-

---

[1] Huygens' *Cosmothereos*, p. 37. [2] *Ibid.* p. 66.
[3] *Ibid.* p. 77. [4] *Ibid.* p. 83.

lar worth and dignity,'[1] nay, would 'venture somewhat farther,' and tell us ' that they have our inventions of the tables of Sines,[2] Logarithms, and Algebra.' And then[3] he argues they must have music, physic, clothing, coaches; must dig metals out of the ground; and probably may manufacture gunpowder; be possessed of mills, engravings, paintings, and practise useful arts, such as printing, glass and clock-making. Nay, would have them making the same discoveries as ourselves, as, for example, of the circulation of the blood. Indeed, in nothing can these imagined inhabitants differ from ourselves, except possibly in shape of body; and even this difference Huygens seems to abandon, and sums up his first chapter in saying all this had been ' proved;' but, thoughtfully considering the reader requires a ' breathing while,' makes a pause.

In the second chapter, or second part of the book, our author visits the several planets, and considers their position, and the condition of their inhabitants; seems to discard the idea of the Sun being inhabited, but peoples the Moon, and affirms that whatever can be said of the Moon may, with very little alteration, be supposed to belong to the attendants of the other planets, passes[4] to the fixed stars, and concludes they are like our Sun, attended by retinues of planets, and finishes by saying:—[5]

'What we allowed the Planets upon the account of our enjoying it, we must likewise grant to all those Planets that surround that prodigious number of Suns; they must have their plants and animals, nay, and their rational ones too, and those as great admirers and as diligent observers of the Heavens as ourselves; and must consequently enjoy whatsoever is subservient to and requisite for such knowledge.'

[1] Huygens' *Cosmothereos*, p. 84.   [2] *Ibid.* p. 85.   [3] *Ibid.* p. 86.
[4] *Ibid.* p. 129.   [5] *Ibid.* p. 150.

Such is the castle, I fear of cards, which Huygens builds up. Truly the inhabitants resting in it 'must be' light as phantasmagorial shadows painted by the magic glass. We had imagined people in dreamy thought half-consciously conceiving such notions; but here they are gravely realized, and reduced to the pleasing assurance of print. Applying his own words in reference to suppositions of Descartes, we may well 'wonder how an ingenious man could spend all that pains in making such fancies hang together;'[1] and yet we are told that the astronomer Flamstead, a contemporary of Huygens recommended the *Cosmothereos* to Dr. Plumer, Archdeacon of Rochester, who was so pleased with it that he left by his will £1800 to found the Plumian Professorship of Astronomy and Experimental Philosophy at Cambridge!

Lalande, from whom I have already quoted, expresses his views in favour of the plurality in a passage which I also quote, both for the sake of showing the grounds of the opinion he maintained, and as containing a mention of some of the eminent men who, subsequently to Huygens, had up to the time (1802) he, Lalande, wrote, discussed the subject. It will be seen (and perhaps in such discussions the precision of logical reasoning is not always perceptible) that Lalande proceeds to a certain extent on views of his own:—

'The resemblance between the Earth and the other planets,' he says, 'is so striking, *that if we allow the Earth to have been formed for habitation, we cannot deny that the planets were made for the same purpose;* for if there is, in the nature of things, a connection between the Earth and the men who inhabit it, a similar connection must

[1] Huygens' *Cosmothereos*, p. 160.

exist between the planets and beings who inhabit them.

'We see six planets around the Sun, the Earth is the third; they all move in elliptical orbits; they have all a rotatory motion, like the Earth, as well as spots, irregularities, mountains. Some of them have satellites, the Earth has one satellite: Jupiter is flattened like our world; in short, there is every possible resemblance between the planets and the Earth: is it, then, rational to suppose the existence of living and thinking beings is confined to the Earth? From what is such a privilege derived but the grovelling minds of persons who can never rise above the objects of their immediate sensations. Lambert believed that even the comets were inhabited (*Système du Monde*, Bouillon 1770). Buffon determines the period when each planet became habitable, and when it will cease to be so from this refrigeration (*Supplemens*, in 4to, tom. 11). What I have said of planets that turn round the Sun, will naturally extend to all the planetary systems which environ the fixed stars; every star being an immoveable and luminous body, having light in itself, may properly be compared with our own Sun. We must conclude that if our Sun serves to attract and lighten the planets which surround it, the fixed stars have the same use. It is thought that the Sun and fixed stars are uninhabitable, because they are composed of fire; yet M. Knight, in a work written to explain all the phenomena of nature by attraction and repulsion, endeavours to prove that the Sun and stars may be habitable worlds, and that the people in them may possibly suffer from extreme cold. M. Herschel likewise thinks the Sun is inhabited.—*Philos. Trans.* 179, p. 155 et suiv.

'Some timid superstitious writers have reprobated this system as contrary to religion: they little knew how to promote the glory of their Creator. If the immensity of His works announce His power, can any idea be more calculated than this to exhibit their magnificence and sublimity? We see with the naked eye several thousands of stars; in every part of the firmament we discover with telescopes, innumerable others; with more perfect telescopes we still find a multitude more. We compute from the number seen through Herschel's telescope in one region of the sky, that there are a hundred millions. Imagination pierces beyond the extent of vision, beholding multitudes of unknown worlds infinitely more in number than those which are visible to our sight, and ranges unrestrained in the boundless space of creation.

'Our only difficulty with respect to the inhabitants of so many millions of planets is the obscurity of the final causes, which it is difficult to admit when we see into what errors the greatest philosophers have fallen; for instance, Fermat, Leibnitz, Maupertius, etc., in attempting to employ these final causes or metaphysical suppositions of imagined relations between effects that we see and the causes we assign them, or the ends for which we believe them to exist.

'If the plurality of worlds be admitted without difficulty; if the planets are believed to be inhabited, it is because the Earth is considered merely as a habitation for man, from which it is inferred *that were the planets uninhabited, they would be useless: but I will venture to assert that such a mode of reasoning is confined, unphilosophic, and at the same time presumptuous.* What are we in comparison of the Universe? Do we know the extent, the properties, the destination, and the connections of nature? Is our existence, formed as we are, of a few frail atoms, to be considered anything when we think of the greatness of the whole? Can we add to the perfection and grandeur of the Universe? These ideas are expressed by Saussure, who, in speaking of a traveller to Mont Blanc, says: If during his meditations the thought of the insignificant beings that move on the face of the Earth offers itself to his mind, if he compares their duration with the grand epochs of nature, how great will be his astonishment that man, occupying so small a space, existing so short a time, can ever imagine *that his being is the only end for which the Universe was created!*

'From these considerations D'Alembert, in the *Encyclopedia* (article "World"), after examining the arguments for supposing the planets inhabited, concludes by saying *the subject is enveloped in total obscurity.* But Buffon affirms that wherever there is a certain degree of heat, the motion produces organized beings; we need not inquire in what way, but imagine these to be the inhabitants of the planets: if that should be the case, we may conclude it highly probable that they are inhabited, notwithstanding the preceding objections.'

Sir William Herschel is referred to by Lalande. He had adduced in a paper on the 'Nature and Construction of the Sun and Fixed Stars,' arguments in favour of the habitability of both Sun and Moon. 'But,' says his biographer, Holden, writing in 1881, 'they rest more on the metaphysical than on scientific basis, and are to-day justly forgotten,'[1] an observation which applies in truth to much which has been quoted from the authors already named.

La Place seems to have taken a view similar to that of Lalande, but in a less reverent, if not indeed in a sceptical spirit.

'Seduced,' he says, 'by the illusions of sense and by self-conceit, we have considered ourselves as the centre of these motions. We imagine, forsooth, that this is for us, and that the stars influence our destinies. But the labours of ages have convinced us of our error, and we find ourselves in an insignificant planet, almost imperceptible in the immensity of space. Inhabitants of this peppercorn, we think ourselves the peculiar favourites of Heaven; nay, the chief objects of care to a Being the Maker of all '—and so forth.

It was to meet infidel views pointing in this direction

[1] *Herschel's Life and Works*, by E. S. Holden, p. 149.

that Dr. Chalmers in 1816 preached, and soon after published, his famous astronomical discourses. Advancing in them reasons for the plurality of worlds, he goes beyond all the requirements of his argument to combat the presumption that Christianity was designed 'for the single benefit of this world.' Whether Dr. Chalmers was warranted in suggesting a proposition which seems to pass the limits of all reasonable speculation, captivating the ear and the mind by the burning eloquence in which it was expressed, may well be questioned. But the arguments used for the actual plurality, although adding nothing to what had been already advanced, state the case perhaps better than they were ever stated previously, and on that account, were it on no other, fall to be given. They are contained in his first discourse.' After referring to the magnitude of the heavenly bodies, he thus proceeds:—[1]

'Now, what is the fair and obvious presumption? The world in which we live is a round ball of a determined magnitude, and occupies its own place in the firmament. But when we explore the unlimited tracts of that space, which is everywhere around us, we meet with other balls of equal or superior magnitude, and from which our Earth would either be invisible or appear as small as any of those twinkling stars which are seen in the canopy of Heaven. Why then suppose that this little spot, little at least in the immensity which surrounds it, should be the exclusive abode of life and of intelligence? What reason to think that those mightier globes which roll in other parts of creation, and which we have discovered to be worlds in magnitude, are not also worlds in use and dignity? Why should we think that the Great Architect of nature, supreme in wisdom, as He is in power, would call these stately mansions into existence and leave them unoccupied? When we cast our eye over the broad sea

[1] *Astronomical Discourses*, Works, vol. vii. p. 21.

and look at the country on the other side, we see nothing but the blue land stretching obscurely over the distant horizon. We are too far away to perceive the richness of its scenery, or to hear the sound of its population. Why not extend this principle to the still more distant parts of the Universe? What though, from this remote point of observation, we can see nothing but the naked roundness of yon planetary orbs? Are we therefore to say that they are so many vast and unpeopled solitudes; that desolation reigns in every part of the Universe but ours; that the whole energy of the divine attributes is expended on one insignificant corner of these mighty works; and that to this Earth alone belongs the bloom of vegetation, or the blessedness of life, or the dignity of rational and immortal existence? But this is not all. We have something more than the mere magnitude of the planets to allege in favour of the idea that they are inhabited. We know that this Earth turns round upon itself; and we observe that all those celestial bodies which are accessible to such an observation have the same movement. We know that the Earth performs a yearly revolution round the Sun; and we can detect in all the planets which compose our system a revolution of the same kind, and under the same circumstances. They have the same succession of day and night. They have the same agreeable vicissitude of the seasons. To them light and darkness succeed each other; and the gaiety of summer is followed by the dreariness of winter. To each of them the Heavens present as varied and magnificent a spectacle; and this Earth, the encompassing of which would require the labour of years from one of its puny inhabitants, is but one of the lesser lights which sparkle in their firmament. To them, as well as to us, has God divided the light from the darkness, and He has called the light day, and the darkness He has called night. He has said, Let there be light in the firmament of these heavens, to divide the day from the night; and let them be for signs, and for seasons, and for days, and for years; and let them be for lights in the firmament of heaven to give light upon their earth; and it was so. And God has also made them great lights. To all of them he has given the Sun to rule the day; and to many of them has He given Moons to rule the night. To them He has made the stars also. And God has set them in the firmament of heaven to give light upon their Earth; and to rule over the day, and over the night, and to divide the light from the darkness; and God has seen that it was good.

'In all these greater arrangements of divine wisdom we can see that God has done the same things for the accommodation of the planets that He has done for the Earth which we inhabit, and shall we say that the resemblance stops here, because we are not in a situation to observe it? Shall we say that this scene of magnificence has been called into being merely for the amusement of a few astronomers? Shall we measure the counsels of Heaven by the narrow impotence of the human faculties? Or conceive that silence and solitude reign throughout the mighty empire of nature; that the greater part of creation is an empty parade; and that not a worshipper of the Divinity is to be found through the wide extent of yon vast and immeasurable regions?

'It lends a delightful confirmation to the argument, when, from the growing perfection of our instruments, we can discover a new point of resemblance between our Earth and the other bodies of the planetary system. It is now ascertained, not merely that all of them have their day and night, and that all of them have their vicissitudes of seasons, and that some of them have their Moons to rule their night and alleviate the darkness of it;—we can see of one that its surface rises into inequalities, that it swells into mountains, and stretches into valleys; of another, that it is surrounded by an atmosphere which may support the respiration of animals; of a third, that clouds are formed and suspended over it, which may minister to it all the bloom and luxuriance of vegetation; and of a fourth, that a white colour spreads over its northern regions as its winter advances, and that on the approach of summer this whiteness is dissipated,—giving room to suppose that the element of water abounds in it, that it rises by evaporation into its atmosphere, that it freezes upon the application of cold, that it is precipitated in the form of snow, that it covers the ground with a fleecy mantle, which melts away from the heat of a more vertical Sun; and that other worlds bear a resemblance to our own in the same yearly round of beneficent and interesting changes.

'Who shall assign a limit to the discoveries of future ages? Who can prescribe to Science her boundaries, or restrain the active and insatiable curiosity of man within the circle of his present acquirements? We may guess with plausibility what we cannot anticipate with confidence. The day may yet be coming when our instruments of observation shall be inconceivably more powerful. They may ascertain still more decisive points of resemblance. They may resolve the same question by the evidence of sense, which is now so abundantly convincing by the evidence of analogy. They may lay open to us the unquestionable vestiges of art and industry and intelligence. We may see summer throwing its green mantle over these mighty tracts, and we may see them left naked and colourless after the flush of vegetation has disappeared. In the progress of years or of centuries we may trace the hand of cultivation spreading a new aspect over some portion of a planetary surface. Perhaps some large city, the metropolis of a mighty empire, may expand into a visible spot by the powers of some future telescope. Perhaps the glass of some observer in a distant age may enable him to construct the map of another world, and to lay down the surface of it in all its minute and topical varieties.[1] But there is no end of conjecture; and to the men of other times we leave the full assurance of what we can assert with the highest probability, that yon planetary orbs are so many worlds, that they teem with life, and that the Mighty Being who presides in high

[1] This anticipation in the case of the Moon, and, it may be said, also in that of Mars, has been realized.

authority over this scene of grandeur and astonishment has there planted the worshippers of His glory.'

I do not pause to examine the correctness of the learned divine's statements, which contain assumptions, or at least suppositions, of fact, upon which 'instruments of observation more powerful' have since borne their evidence or shed a light. I will not anticipate in which direction, but I may observe that underlying all this burst of noble eloquence is the idea that worlds can only be created for habitation. Farther on he advocates the view that each of the 'stars may be the token of a system as vast and as splendid as the one we inhabit. Worlds roll on these distant regions; and these worlds must be the mansions of life and of intelligence.'[1]

It is not surprising that the great preacher should have been assailed for the positions he had thus taken; but, so far as I have seen, his assailants were foes unworthy of his steel, and he does not appear to have deigned to notice them. He was attacked principally upon the religious view of the question; the subject of the plurality itself being all but ignored, and scarcely discussed. One writer, Alexander Maxwell, in a book of 265 pages,—an elaborate series of prosy 'letters, notes, and memoranda, philosophical and critical,' occasioned by these discourses,—setting out by disputing astronomical verities, and telling us how much he was indebted to the study of mathematics under one of the first mathematicians of the age, argued with a great want of dialectic pertinence, and with much semblance of learning, to very little effect. Disputing as he does the doctrine of plurality, he advances nothing of any moment bearing on the question.

[1] *Astronomical Discourses*, Works, vol. vii. p. 31.

After this the subject seems to have slept for nearly twenty years. No one had ventured boldly into the arena to contend, on adequate grounds, for the non-plurality.

But in 1837, Dr. Thomas Dick, originally a minister of the Scotch Secession Church, who abandoned the clerical profession to teach, to lecture, and to write popular works, published his *Celestial Scenery*, descriptive mainly of the bodies composing the solar system, yet devoting a considerable portion to the enforcement of the doctrine of plurality of worlds. His books, which contain much information, can scarcely be called authoritative; but, so far as I have observed, *Celestial Scenery* seems to have been, in point of time, the next work in which the subject is considered at any length. It enters upon the question in a novel manner; for its first office in the case of each body is to settle the number of its population, which is done by estimating the total extent of superficial area, and taking as the standard of calculation the population of England, reckoned at 280 inhabitants for every square mile. Thus doing, the population of each planet and its satellites is calculated by the multiple of its square mileage. As if to make it as absurd as possible, no deduction is made for ocean, desert, or ice, and no allowance is granted for difference in size or description of the inhabitants presumed to exist upon each. The grand total, it readily may be imagined, is astonishing. The results are gathered together in a table interesting to the curious, and persuasive to the credulous:—

Planets.	Population.
'Mercury,	8,960,000,000
Venus,	53,500,000,000
Mars,	15,500,000,000
Vesta,	64,000,000
Juno,	1,786,000,000
Ceres,	2,319,962,400
Palles,	4,009,000,000
Jupiter,	6,967,520,000,000
Saturn,	5,488,000,000,000
Outer ring of Saturn,	
Inner ring,	8,141,963,826,080
Edges of the ring,	
Uranus,	1,077,568,800,000
The Moon,	4,200,000,000
Satellites of Jupiter,	26,673,000,000
Satellites of Saturn,	55,417,824,000
Satellites of Uranus,	47,500,992,000
Amount,	21,894,974,404,480.'[1]

Dr. Dick cannot be blamed for the omission of the planets subsequently discovered. Neptune and his satellites, and the 150 additional planetoids, besides the little moons of Mars, would, with all their populations, have considerably swollen the splendid total to which, without them, the enthusiastic doctor had thus, with ever-to-be-praised solicitude, attained. Elsewhere he reckons the population of the Earth at a modest 800,000,000, and of the Sun (which, though not positive on the subject, he[2] evidently leans to holding inhabited) at 681,184,000,000,000, or equal to the inhabitants of 850,000 worlds such as our own. Indeed, it will be observed that, with the single exception of the small planetoid Vesta, every one of the bodies is credited with a population greater, and in some cases many thousand times greater, than that of the Earth. Even our Moon receives more than five times the population of the planet upon which it humbly

[1] *Celestial Scenery*, p. 285.   [2] *Ibid.* pp. 219. 220.

waits. The calculation is truly wonderful. But when he was so agreeably engaged, it is surprising he did not study arithmetical exactness to the extent of condescending to detail the hundreds and the units; and as there must in all cases be many, in all of the planets, who are under age,—infants that travel half-price,—why did he not both serenely estimate the units, and secure that one at least should honourably ride off with an odd half? Dean Swift, with that feeling attention to the exigencies of vraisemblance for which he was distinguished, could scarcely have omitted the odd half; nor would Defoe.

Starting with these magnificent visions of planetary population, he[1] devotes a considerable portion of his book to the demonstration of the plurality of worlds. His arguments are these:—

1. There are bodies in the planetary system of such magnitudes as to afford ample scope for myriads of inhabitants. It would be a very gloomy view of the Creator to imagine them barren and desolate.

2. There is a general similarity[2] among the planetary bodies, tending to prove that they were all intended to subserve the same ultimate designs. They are all (1) spherical, (2) solid. (3) They all revolve round the Sun, (4) and on their axis. They are all (5) opaque bodies deriving their lustre from the Sun, (6) kept together by the law of gravity round the central luminary, obeying the same laws and subjected to the same influences; so that the reasonable conclusion is, that the ultimate destination is the same, and that they are all replenished with inhabitants.

3. There are special arrangements, indicating adapta-

[1] *Celestial Scenery*, p. 342.   [2] *Ibid.* p. 342.

tion to the enjoyment of sensitive and intelligent beings, proving this to be the ultimate design of their creation;[1] because they are (1) diversified into hill and dale; (2) environed with atmosphere; (3) provided with means of distribution of light, heat, and colour; (4) provided with moons; (5) adjusted by comparative density and rapidity of rotation to their respective sizes—to suit organized intelligences.

On the assumption of the truth of the statements thus made, Dr. Dick argues that God has a design in view in all His arrangements; that whenever we find a contrivance exactly adapted to accomplish a given end, we may be sure that was the end to be accomplished; that we see in all the planets special contrivances calculated to promote the enjoyment of myriads of intelligent agents; and, as God is possessed of infinite wisdom, to suppose them uninhabited would be unworthy of infinite wisdom, and 'the thought would be impious, blasphemous, and absurd:'—a conclusion, I must observe, which may perhaps be reached very logically from the premises, provided always the premises be, as matter of fact, true. But are they so? We shall see.

4. The scenery of the Heavens, as viewed from the surfaces of the larger planets and their satellites, forms a presumptive proof they are inhabited by intellectual beings; and the reverend doctor[2] tells us how much grander the views obtained from these satellites are than any we poor inhabitants of Earth enjoy; and concludes it is impossible to suppose that such magnificent views would be unless there were rational beings capable of appreciating them. The proper corollary from which conclusion, I suppose, is that the inhabitants of our Moon must be many degrees more intellectual than the

[1] *Celestial Scenery*, p. 353.   [2] *Ibid.* p. 364.

men of Earth; and still more should the favoured inhabitants of Jupiter's moons, whose views, we take it, are so transcendently more magnificent.

5. Every part of nature in this world, he says,[1] has been destined to be the support of animated beings, and therefore it would be absurd to suppose the planets to be destitute of life. Matter was evidently formed for mind; and as God delights in conferring existence and happiness on every order of beings, it must be in accordance with our conceptions of the Divine Being to suppose that the planetary regions are peopled, and with beings beyond doubt possessed of rational and intellectual natures.

Apart from the assertions in point of fact contained in Dr. Dick's statements, his argument, thus epitomised, is pervaded by the general assumption that the orbs must be created for habitation, which, if not implying assertion of the final cause, may be said to be the *de quo queritur*.

In his *Sidereal Heavens*,[2] published subsequently, after setting out in a separate chapter his view that the stars were constructed as centres of planetary systems, he enters, very much on the same lines,[3] upon a further argument for the plurality of worlds, drawn chiefly from his suppositions regarding the perfections of the infinite Creator; and he also attempts to prove his position by quotations from Scripture.[4] One chapter is devoted to describing the inhabitants, very much on Huygens' plan of resembling them to men.

It would not seem that either work produced any visible movement, or evoked any fresh literature. If

[1] *Celestial Scenery*, p. 366.   [2] P. 208, c. 14.
[3] C. 16, p. 234.   [4] C. 17, p. 252.

the slumber was broken, it was but a drowsy wakening to listen half-consciously to the reverend divine's cogitations, which only squared with preconceived views; for, indeed, the plurality was then tacitly assumed.

But not long after there appeared a notable work, the *Vestiges of the Natural History of Creation,* wherein the unknown writer went a step farther, and advanced the position that organic beings are produced or evolved in all the orbs by virtue of a general law of nature, and out of an original fiat of the divine will, by which—

'The whole train of animated beings are to be regarded as a series of advances of the principle of development;'[1] and so, the author informs us, 'organic life presses in wherever there is room and encouragement for it, the former being always such as meet the circumstances.'[2] 'We have to suppose that any one of these numberless globes is either a theatre of organic being, or on the way of becoming so.'[3] 'Inorganic matter must be presumed to be everywhere the same with differences in the proportions of ingredients in different globes, and also some difference of conditions.'[4] He then assumes that 'where there is light there will be eyes, and these in other spheres will be the same in all respects as the eyes of tellurian animals, with only such differences as may be necessary to accord with minor peculiarities of condition and of situation. It is but a small stretch of the argument to suppose that one conspicuous organ of a large portion of our animal kingdom being thus universal, a parity in all the other organs—species for species, class for class, kingdom for kingdom—is highly likely, and that thus the inhabitants of all the other globes of space bear not only a general, but a particular resemblance to those of our own.'[5]

[1] *Vestiges of Creation,* 4th ed. p. 208.　　[2] *Ibid.* p. 164.
[3] *Ibid.* p. 165.　　[4] *Ibid.* p. 167.　　[5] *Ibid.* p. 168.

Taking such views, the inevitable conclusion to which this author was led, was that organic beings have 'all come into existence by the operation of laws everywhere applicable.'[1]

The literal meaning of the book was simply this, that God having once for all spoken His will and put the wheels of nature in motion, for ever afterwards restrained His hand. The author could not imagine the possibility of God exercising the creative power at each successive new form of life, and so, because the ways of Him who letteth not a sparrow fall on the ground without His will were incomprehensible to the imperfect knowledge and finite powers of the feeble creature, the thought was to be regarded as taking 'a very mean view' of Him—it was, in short, 'ridiculous.' One thing, however, is observable, that after endeavouring to trace the successive steps of change from one condition of being to the immediately higher,—as if the fact of resemblance in the successions proved their descent the one from the other, or proved it to happen without the interposition of divine direction and divine power,—he stops short of man at the quadrumana, and concludes by saying, 'Here such obscurity prevails that I must be content to leave the task to other inquirers.'[2]

The author of the *Vestiges* was by no means the first to suggest the view of development, for he had been preceded by several writers who raised the question; but the boldness of his views and the dogmatic vigour of his style produced a sensation and an influence which probably had not been previously experienced, and he may be said, therefore, practically to have given birth to the doctrine of evolution now so popular among many naturalists, and with which we shall have after-

---

[1] *Vestiges of Creation*, p. 169.   [2] *Ibid.* p. 272.

wards more particularly to deal. The *Vestiges* was first published in 1844, to which was added a sequel of *Explanations* in 1846, and it rapidly passed through ten editions, in which there are numerous alterations and excisions apparently indicating from time to time some slight modifications or some slight developments of its author's daring opinions. It experienced a wide criticism, and it is to be feared the answers made were not always on the correct lines.

Hitherto, it will be seen, although all would not have indorsed the views of the *Vestiges*, the opinion of every one of those to whose writings I have adverted had been in favour of plurality. About 1853 or 1854, however, there appeared a book, called *The Plurality of Worlds: an Essay*,—anonymously though speedily and, as it proved, rightly attributed to the Rev. Dr. William Whewell,[1] Master of Trinity College, Cambridge, and author of the Bridgewater Treatise, *Astronomy and General Physics considered with reference to Natural Theology*. In the Bridgewater Treatise, while discoursing of 'the vastness of the Universe,' Dr. Whewell

[1] In *The Life and Selections from the Correspondence of William Whewell, D.D.*, by Mrs. Stair Douglas (1881), there occurs this incidental acknowledgment of the authorship in a letter 'to Kate Marshall,' dated 2d January 1854 (p. 433): 'There is a little, not mystery, for that I hate, but reserve to be used in talking of it as my book, for it is published without my name, and contains notions that may be startling to some persons, though I am persuaded that they tend to give us a true view of God's government of the world. I suppose, after all, you must have heard me talk of a book about *The Plurality of Worlds*, which I was writing or finishing when you were here, and now the murder (of the inhabitants of Jupiter) is out.' This *Life*, which is nearly altogether made up of letters, is disappointing. There is no direct mention of the publication either of the essay or of the treatise, nor any mention of the reception of either by the public, notwithstanding the amount of discussion to which the essay especially gave rise.

incidentally observed that 'no one can resist the temptation to conjecture that these globes' (the planets and the satellites), 'some of them much larger than our own, are not dead and barren; that they are, like ours, occupied with organization, life, intelligence.' But in the Essay he entered the lists as a champion to combat that view, and so far as I have observed he seems to have been the first who did so, or did so with any force of reason.[1] For he discusses the whole subject fully and systematically, though not always so as to command concurrence, indeed also with all the overlading amplitude of a theologian. In the early chapters he sets out the difficulties which had originated the famous *Astronomical Discourses* of Dr. Chalmers, gives due weight to the doctor's argument, and discusses the subject. He then proceeds to point out from geology the antiquity of life upon the Earth as compared with the recent creation of man, and considerations to which this fact gives rise. Thence he proceeds to the stars and planets, and argues from scientific facts against the conjecture of their being inhabited, and against even the probability of the fixed stars being, at least in every case, surrounded by planetary systems, because, among

---

[1] Such *obiter dicta* as those of John Wesley, quoted in Powell's *Unity of Worlds*, p. 302, can hardly be considered, in this collocation, as offering an exception. Wesley's words, however, are remarkable: 'Suppose there were millions of worlds, yet God may see in the abyss of His infinite wisdom reasons which do not appear to us why He saw good to show this mercy to ours in preference to thousands or millions of other worlds. I speak this even upon the common supposition of the plurality of worlds, a very favourite notion with all those who deny the Christian revelation, and for this reason, because it affords them a foundation for so plausible an objection to it. But the more I consider that supposition, the more I doubt of it, insomuch *that if it were allowed by all the philosophers in Europe, still I could not allow it without stronger proof than any I have met with yet.*'

other reasons, of the large proportion of double stars. He argues from the advance of our planet to a peculiar condition and situation in a temperate zone, and the high nature and destiny of man to the Earth being the special seat of God's peculiar care, and that man has been placed upon it as a preparatory stage to fit him for eternity, and that the Universe is subordinate to man; while he combats the idea that the refusal to acknowledge inhabitants in other stellar bodies is adverse to the thought of the goodness, benevolence, or majesty of God. It was a thoughtful, able work, although many of the views it contains, such as the assumption that the Universe was created for man, are unnecessary to the conclusion, and are far from carrying conviction. It excited so much attention, as within a year to run to a fourth edition.

The Essay stirred up discussion. Apart from the observations and comments of reviewers, he was attacked by two writers. One of them anonymously assailed him in a small book, called *The Plurality of Worlds: The Positive Argument from Scripture, with Answers to some late Objections from Analogy*. It was a dreary argument, which may be put aside without remark. The other was from the pen of the venerable Sir David Brewster. But the want of temper manifested in Sir David's book, called *More Worlds than One the Creed of the Philosopher and the Hope of the Christian*, disfigures it, and manifests an amount of bias which disqualified him from forming calm and sound conclusions. Viewing the essay 'as an elaborate attack upon opinions consecrated by reason and revelation,' having a 'direct tendency to ridicule and bring into contempt the grand discoveries in sidereal astronomy,' he allows himself to 'reject the ignoble sentiments with indignation,' and hints at the writer being either 'a fool or a madman, writing only from a love of notoriety.' On the other hand, the fantastic views of Fontenelle and Huygens receive unbounded regard; and, indeed, he follows to some extent in their wake, and arguing with the aid of additional scientific facts, or supposed or assumed facts, he easily takes his conclusions, by adopting Huygens' method even to the using of his very words: '*it must be.*' The idea underlying the whole book is that all the orbs must have been made for some use, and that use must have been for the abode of rational beings; that all the stars are suns, each one of which is the centre of a system of planets, and there being one planet, at least, in the solar system inhabited, there must be at least one in each of these assumed systems similarly inhabited. It is rather curious to find that Sir David's dogmatical assertions of astronomical fact are in some instances distinctly set aside by the later discoveries of the spectroscope and otherwise; as, for example, his assertion that all nebulæ are resolvable into stars, that there are volcanoes in the Moon and an atmosphere surrounding it, together with his views regarding the planets and the eternity of the Sun. In the overbearing spirit in which it was written, well-founded argument was not to be expected; and, straining his views as he does, it may almost be said that, in speculating on the plans of the Almighty, he is sometimes, to use his own words applied to the essayist, 'bordering on the blasphemous.'[1] While Whewell laid himself open to attack, I cannot, however, say that Sir David Brewster was remarkably successful in his assault, even when, in common with some other writers, he is armed with words of Scripture, which he vainly wrests in support of his conclusions.

[1] *More Worlds than One*, ed. 1870, p. 229.

A far more philosophical and valuable communication was made at same time by Professor George Wilson, in a small publication of 50 pages, which he called *The Chemistry of the Stars*. Exception might be taken to the fanciful clothing in which some of his ideas were dressed,—to his jury of chandler, stoker, sailor, and the like, and their professional remarks. But the idea which permeates the treatise is to point out the differences, actual and possible, between the Earth and the other orbs of creation, and hence to reason against the plurality of worlds. The argument is exactly in the right direction, and, so far as it goes, is supported by chemical facts and proofs of the most telling character. But the knowledge then possessed was more limited than it is now; and it is only much to be deplored that the talented popular chemist did not survive to carry his demonstration to a wider field, to do which no philosopher was more competent. His tract afforded, so far as I have seen, the most important support to the position taken by Whewell.

On the other hand, Mr. Montague Lyon Phillips entered the lists to champion the side of plurality. His book, *Worlds beyond the Earth*, is mainly occupied with a demonstration of the Nebular hypothesis. But he has not been so successful in his attempt to prove the plurality. He is forced to admit that the Sun, Moon, and planets exist in conditions differing from those of the Earth. The chemical substances—the gases, the fluids, the atmospheres—are different; and, driven to admit that human beings could not exist on those other bodies, he only finds his way out of the difficulty by suggesting that these inhabitants must be differently constituted. How far this is an admissible position in the present question may afterwards appear.

The Rev. Baden Powell, in his *Unity of Worlds*, took up the parable, and, with a show of impartiality, disclaiming controversialism, hitting both sides, he not the less turns the scale for the plurality, to his own satisfaction, by quietly placing the weight of his foot in the balance. Referring to the discussions on the Nebular hypothesis, he pointed out that there was no necessary connection between that hypothesis and the question of plurality, although he leans to it, and supposes a common origin for all the planets, and that all are suffused with the same material (a fact which Phillips did not admit); and, discussing the views relative to the Sun, Moon, and planets, seems to arrive at the conclusion that they are of the same kind or character, but only differ in degree; and dismisses the geological difficulty of man's recent appearance by assuming a theory of progressive development or preparation; so that if a world is not at present inhabited, it is because it has not reached, or has passed, the condition of fitness for habitation. He also at great length discusses the arguments maintained on both sides from the theological point of view, maintains that it was a mistake to remove the subject from the region of inductive conjecture to that of supporting theological belief, and condemns the impropriety of attempting to solve a philosophical problem on any ground but physical analogy. In doing so he hits both sides,—hits Sir David Brewster, who starts with the assumption that the worlds were made for the sole purpose of being inhabited,—which was just begging the question, and an assumption of final causes not warranted by sound principles of reasoning. But then Whewell, in the essay, he thinks, was equally wrong in seeming to advance (for it is rather a gloss upon his argument) that the rest of creation must be supposed a waste in order to dignify man. The argument from assumed final causes was thus, he says, applied with equal force for opposite conclusions. It was likewise a mistake, which some had made, to deny the existence of inhabitants in other worlds, lest it should clash with our thoughts regarding man, and revelation, and the sacrifice of the Son of God for man—all this was to disregard the proper province of theology. But though thus apparently reproving all, there is an under-current pervading the section, indicating his favour for the subordination of man and the peopling of other worlds.

The controversy thus raised by Whewell's essay brought some other writers into the field; but it seems to have died down, and apparently was resuscitated in France, in 1862, by the French astronomer Flammarion, in the publication of his popular work, '*La Pluralité des Mondes Habités;* Etude ou l'on expose les conditions d'habitabilité des terres célestes discuteés au point de vue de l'astronomie de la physiologie et de la philosophie naturelle,' written, he tells us, for the purpose of demonstrating that the plurality of worlds is a doctrine, at once scientific, philosophical, and religious, of the highest importance. He followed it up in 1865 by another, '*Les Mondes Imaginaires et les Mondes Réels*, voyage pittoresque dans le ciel et revue critique des théories humaines scientifiques et Romanesques, anciennes et modernes sur les habitants des astres.' This work, which is for the most part occupied with a very detailed history of the theories, ancient and modern, regarding the inhabitants of the stars, is in this respect an enlargement of the previous book, and in the last chapter he states that the appearance of his former work was the signal for the revival of attention to the subject. It was rapidly disseminated throughout France and in Europe, and was translated into all the languages of the two continents, and was at the time he wrote in its 21st French edition.[1] It had given birth to many books and the reproduction of others, and he catalogues over thirty, principally French, published since 1862.

Flammarion likewise devoted a chapter of *Les Merveilles Célestes* to the plurality of inhabited worlds. But it hardly appears to me that he added anything to the views which had been advanced by earlier writers, and, indeed, the little he does advance may be summed up in resting his view for plurality on the insignificance of this Earth, the abundance of life upon it, and similarity of this planet to the others.

Meantime the English press had been almost silent. At length, in 1870, Mr. Proctor published *Other Worlds than Ours*, the remark upon which by Flammarion[2] is, 'Aux yeux de tous ses lecteurs, l'auteur n'a évidemment qu'un tort : c'est de paraître ignorer que d'autres écrivains ont traité la même question avant lui, et de la même manière, et de ne citer que des auteurs anglais, comme

---

[1] My copies are dated 1876, and *La Pluralité* is the 24th, the other the 14th edition. The two books run to nearly 1000 pages, and I must confess to have done not more than glance at them. In both works Flammarion has industriously gathered together apparently everything that has been written for or against the plurality, and whether purposely or merely cursorily, whether seriously or in jest. Probably, therefore, those desirous of making themselves fully acquainted with the whole literature of the subject may find the means of study in his pages.

[2] *Mondes Imaginaires*, p. 582.

si l'idée de la pluralité des mondes était née en Angleterre et n'avait été cultivée que dans cette île.'

Mr. Proctor in this treatise, however, proceeds in an orderly manner to examine the conditions in which the various bodies are placed, and his views on each will best be referred to as each comes to be afterwards discussed. Mars, he believes, is the only planet unmistakeably showing signs of life, and he by no means maintains life at present in any of the others. There are conditions in them inimical to life, as, for example, in Jupiter, which 'is still a glowing mass, fluid probably throughout, still bubbling and sending up continually enormous masses of cloud, to be gathered into bands under the influence of the swift rotation of the giant planet.' But Mr. Proctor takes the view that the worlds were created for life, and leans to the supposition that the beings which people other globes are adapted to the conditions of each, and that if not at the present time habitable, these globes have been or may yet become so. He touches on the subject or deals with it in several others of his many works, and particularly in *The Orbs around us*, *Flowers of the Sky*, *The Expanse of Heaven*, and *Our Place among Infinities*, 'a series of essays contrasting our little abode in space and time with the infinities around us.'

In the last-mentioned book he devotes a special chapter to considering 'A new Theory of Life in other Worlds.'[1] Contrasting the views of Whewell and Brewster, he starts by observing that Whewell did good service in breaking the chains of old-fashioned ideas and inaugurating freedom of discussion, and while admitting that the balance of evidence was in favour of Whewell's theory, considers that Brewster's is the one which com-

[1] *Our Place among Infinities*, 1875, p. 48.

mends itself more favourably to the mind which would believe that God hath done all things well, and nothing *that He hath made was made in vain.* And having forced on us the belief that the support of life is the object for which the Earth was created, we are thus 'led to regard the other orbs, which, like her, circle round a central Sun, as intended to be the abode of life.[1] But there was a middle course; and pointing to the short duration of life, and especially rational life upon Earth, as a mere ripple on the great ocean of time, leading to the thought that rational life was but a subordinate object of the Earth's creation,—that there is no ground to suppose that this particular epoch is the epoch for life everywhere, and that the probabilities are that it is only the epoch for a small proportion,—the view he takes is, that the planets of this and other systems are passing through stages, and the stage of life has only been reached or is occurring in a proportion of the planets; but this proportion, over the whole Universe, will amount to millions.

I have not observed that of late years any other English writer has ventured to take the matter up as a separate subject of discussion, although some have dealt with it along with other themes. Thus Dr. Phipson devotes a chapter of his *Mysteries of Nature* to inhabited planets, in the course of which he thus writes:—[2]

'To any person tolerably devoid of prejudice, and slightly acquainted with the teachings of modern astronomy, it must appear absurd in the extreme to suppose that these magnificent worlds which revolve round the Sun should not have been as highly endowed by the Creator as our little Earth,—a mere point in the Universe,

---

[1] P. 57.   [2] Phipson's *Mysteries of Nature*, 1876, p. 268.

—which constitutes so moderate a feature among them. And it may perhaps be quite as fanciful to imagine that our globe is for man the best possible of worlds. One or two prodigious difficulties arise, however, when we wish to bring forward some palpable proof of the planets being in reality inhabited by creatures at all like ourselves. Not the slightest doubt can exist as to the possibility of this, as far as certain planets are concerned, more especially Venus, Mars, and Mercury, reasoning from the little we know of their physical properties and their telescopic appearance, so similar to what our Earth must appear viewed from one of them.'

Dr. Phipson, however, has his doubts about the satellites, the Sun, and the larger planets. How far his views are justified remains to be seen. Professor Newcomb, an American astronomer, devotes a few pages, at the close of his work on astronomy, to the subject. He says:[1] 'When we contemplate the planets are worlds like our own, and the stars as suns, each perhaps with its retinue of attendant planets, the idea naturally suggests itself, that other planets, as well as this, may be the abode of intelligent beings;' and after alluding to the importance attached to the discovery of evidence of life in them, says: 'It is therefore extremely disappointing to learn that the attainment of any direct evidence of such life seems entirely hopeless, so hopeless, indeed, that it has almost ceased to occupy the attention of astronomers,' and observes that it is 'quite possible that retinues of planets revolving in circular orbits may be rare exceptions rather than the rule among the stars;'[2] but granting planets, all the chances are decidedly against the idea that any considerable fraction of the heavenly bodies is fitted to be the abode of such animals as we have on earth, and that the

[1] Newcomb's *Astronomy*, p. 516.   [2] P. 517.

number of those which have the requisites for supporting civilisation is a very small fraction indeed of the whole. This conclusion, he says, rests on the assumption that the conditions of life are the same in other worlds as in our own, and this again on whether there be limits to adaptability on earth; and then states that such limits do exist, and the higher the condition of life the more restricted are the conditions, and if great changes are to occur on the surface of the globe, the higher forms of life would be destroyed. He is therefore led to the conclusion, that 'in view of the immense diversity of conditions which probably prevails in the Universe, it would be only in a few favoured spots we should expect to find any very interesting development of life.' Professor Newcomb therefore adopts a view different from that of Dr. Phipson. But scientific men have been too much engaged within the last twenty years in making investigations with the spectroscope and other astronomical appliances, and thus getting at astronomical facts, to deal with what has always been regarded by astronomers as a pure speculation. The discoveries made by means of spectrum analysis must necessarily have proved, and have truly proved, of greatest consequence, and they tend to lead to more certain conclusions on many points of the question involved. Mr. Proctor seems to think, in one of the chapters of *The Orbs around us*, that the discoveries by this analysis refute some of the arguments or positions of Mr. Whewell; but I have purposely reserved all discussion on the matters they involve to their proper place in the after pages.

II.

SUCH is a sketch of some of the leading stages in the history of the question, and of the views of some of the leading writers. Those who believe in the plurality are not quite agreed regarding the nature of the inhabitants. But their general contention seems to be that these inhabitants are rational beings; and, indeed, it seems difficult to reconcile the theory of life in those other orbs unless upon the footing that the life is rational,[1]—that the occupants are intelligent and capable, like man, of apprehending the existence of the Creator,—for to this the argument naturally leads. Some, like Huygens, have even contended for their resembling man in all respects. Others do not give their ideas this concrete form, and content themselves with simply affirming the habitability of all orbs, believing that the beings which constitute their population must vary in shape and kind in accordance with the constitution of the planets on which they severally dwell, and the varied circumstances in which they are respectively placed. While others again contend merely for the potentiality of the different orbs to receive life, and say that they, or many of them, may only yet be in a state of preparation for life, or that they have ceased to sustain it.

But it will also be noticed that the arguments

[1] See p. 171.

advanced in support of the plurality are various. They may be resolved into several propositions:—

1. It is said that many passages in Scripture imply that the orbs are inhabited. In 1801, the Rev. Edward Nares, Rector of Beddinden, Kent, was at pains to write a whole book of 404 pages, entitled '$EIΣ\ ΘEOΣ\ EIΣ\ MEΣITHΣ$: or, An attempt to show how far the philosophical notion of a plurality of worlds is consistent or not so with the language of the Holy Scriptures;' and it is only fair to say that, while it is a prolix, mild, and inconclusive work, it appears to be honest and, as far as possible, impartial, although one is much tempted to think the writing of it was labour wasted.

In a very different spirit Sir David Brewster devoted a chapter of his *More Worlds than One* to the subject, and to this I would particularly refer.

Without specifying the several passages in detail, I may notice a few. Thus he refers to the words of David flowing from an 'inspiration which no doubt revealed to him the magnitude, the distances, and the final causes of the glorious spheres which fixed his admiration,'[1]—an assumption large enough indeed,— 'when I consider the Heavens, the work of Thy fingers, the Moon and the stars which Thou hast ordained, what is man that Thou art mindful of him, and the son of man that Thou visitest him?'[2] or those of Job: 'The Heavens being spread out as a tent to dwell in;'[3] 'the chambers of the south;' to other writers speaking of the 'host of Heaven' (a well-known synonym for the stars); to Isaiah: 'the Lord who created the Heavens'

---

[1] *More Worlds than One*, p. 10.   [2] *Ibid.* p. 10.
[3] *Ibid.* p. 12.

and the Earth; 'He created it not in vain, He formed it to be inhabited' (a passage which, literally taken, would imply that the *Earth* only was formed to be inhabited); or to the New Testament in the use of such words as 'worlds,' 'the Heavens,' or 'all things.' But if other similar passages of Scripture were taken literally, what should we make of such words as these: 'The stars shall fall from Heaven;' 'The stars of Heaven fell on the Earth;' 'Praise Him, Sun and all ye stars of light'? It is in vain to take such words literally, and to catch at a hidden meaning of this nature from the fervid utterances of the poet or the prophet, while it is manifest that the passages in question neither directly affirm that the stars are inhabited,—and, indeed, it is a proposition now all but abandoned that the stars themselves are so, the contention being merely that their supposed but invisible planetary attendants are,—nor can they, not one of them, sanction the deduction of such a meaning. The Bible was not given for the purpose of teaching us scientific facts, nor does it. It is only the extravagant fancy of a keen and inexact advocate which can torture undesigned expression to suit his argument in a way which will not brook the smallest investigation or criticism. It is a method of argument to be decried and dismissed.

2. It is maintained that the supposition of the orbs being inhabited redounds to the glory, and manifests the wisdom of God; and the contrary supposition would be to lessen the glory and impeach the wisdom of God. But this is truly to affirm that it consists with the knowledge of those who so aver that the glory and wisdom of God can only be manifested in the other orbs by peopling them; or to declare—what it has not been given to man to know—wherein the glory and wisdom

of God consist. It is, however, just another form of saying—

3. That the orbs were all made for the express purpose of being inhabited, and would be useless unless they were, which, besides being a begging of the question, is an assumption of the final cause, for which it cannot be pretended we have the smallest warrant.

The idea in this assumption truly lies at the bottom of nearly all that has been said in support of the plurality. Looked fairly in the face, it cannot stand.

For it must be acknowledged that we are all finite beings. We know very little, even of the Earth on which we live, and we are apt to measure everything by our own narrow conceptions and limited experiences, and to imagine there can be nothing existing different from what we have seen or heard. But this we do know, that everywhere, so far as it comes within our observation, the Great Creator has manifested His almighty power by a most wondrous diversity of gifts and operations. We find in endless multiplicity varied forms of matter, of animal and of vegetable life, in the different regions of the globe. There are no two countries, or even two districts of a country, alike. There are no two mountains, or valleys, or rivers, or trees alike. And passing from this Earth, 'there is one glory of the Sun, and another glory of the Moon, and another glory of the stars; for one star differeth from another star in glory.' Nay, even in things which are most alike is there special difference. No two men or women are alike; no two sheep are alike, for the shepherd can tell them all by name, and the ewe knows its own lamb among the largest flock.

We should therefore be led rather to suppose that, while

the allwise and bountiful Creator has seen fit to bestow upon this world the gift and adornment and the glory of life as its peculiar heritage, He has, out of His immeasurable treasury, imparted to the other orbs each a separate blessing in special benefactions to which our experience does not reach, and of which our finite minds can form no conception. It may be so. We cannot tell. In this life at least it must, without a special revelation, be from us for ever hid.

Or it may be that He, in the plenitude of His sublimity, has created them without a use, and that He delights in them simply for their beauty and their perfection, and as a manifestation of His glory. It may be so. We cannot tell. But He has given to us the glowing opal and the lustrous diamond; and these may be to man in themselves useless, apart from their soft light or their brilliant radiance, and yet for this useless glitter they may to him have priceless value. And may not He who covers Himself with light as with a garment, and stretches Himself, through varied scenes of grandeur inconceivable, to infinity, like 'as a bride adorneth herself with her jewels,' delight to crown His royal head with a universe of sparkling orbs? Or as we, finite beings, made in His image and given to share somewhat in His thoughts, can never tire of looking up to the star-studded sky and drinking unbounded pleasure from the mere contemplation of those far-off worlds, dwindled down in the distance, in our puny sight, to specks of light upon the firmament; or as master spirits among men regard with complacent satisfaction the works which their own hands have designed or reared,—so may He, with His all-seeing eye—to which nothing in all its grandeur or in all its loveliness is lost—behold with a majestic serenity, to which we cannot soar, the stupendous fabric which His omnipotence has 'spread out like a curtain' in all its circling wonders,—where the great is perceived in all its magnitude and the small is not obscured or hidden by the vast,—created in all that amazing perfection which constitutes to our minds the matchless beauty of nature,—and bend to it that approbation which, when looking on it at the first, we are told in the simple words of the inspired creation hymn, 'God saw that it was good.' Can we not realize that these, His works, made in the fulness of that holy exaltation in which they were conceived, have throughout eternity brought to that glorious Being the joy unspeakable which fills His presence and abides with Him for evermore? For it is written, 'Thou hast created all things, and for *Thy pleasure* they are and were created.'

But when the argument leads, by a necessary sequence, to speculations like these, we feel that we are passing beyond the region of legitimate reasoning; for we are moving into depths we cannot fathom. We can drop no anchor there. Arguments which rest on conjectures regarding final causes can lead to no result. The view, therefore, which has been propounded, must be disregarded as worthless.

With this reason there are several others which must also be cast aside as involved in it, or as just other ways of putting the same idea. Among them the following may be placed:—

4. That every part of the World is full of life, therefore it would be absurd to say that the other planets are destitute of life.[1]

[1] It may be noted that the Earth has not throughout its whole career borne life, and that, as we shall afterwards see, the period during which

OTHER REASONS CONSIDERED. 139

5. The finger of God and the wisdom of the divine government are more clearly manifested in the supposition of endowing the other orbs with life. Are we to deprive the other planets of plants and animals which, more plainly than inanimate matter, speak the divine Architect?

6. That to give life alone to the Earth, an insignificant body on the Universe, is to subordinate all the others.

7. The Earth having been made for habitation, the other orbs must be presumed to have been so too; although, perhaps, these other orbs, or some of them, have not yet reached the stage fitting them to receive life.

8. That life is the proper object for which these orbs were created, the use for which they were made.

These propositions, after what has been said, require no comment or answer; for the reasons why the third proposition cannot be legitimately maintained apply to and cover them also.

But it has been suggested—

9. That the peopling of this world solely is to suppose that the whole remaining creation was made for the gratification of man.

Now, though it be evident that all things on this Earth were subordinated to man created in the image of God, and that to him was given 'dominion over the fish of the sea and over the fowl of the air, and over every living thing that moveth upon the Earth,' and that to him was also given 'every herb bearing seed

---

it has been fitted for man's occupation has been only a speck or point in its history. The future duration of this condition may possibly not increase materially the magnitude of the point or speck.

which is upon the face of all the Earth, and every tree in the which is the fruit of a tree yielding seed,' to 'be for meat,'—that the Earth has been adapted in many ways, and adjusted in the system to which it belongs, to support his life,—it cannot be said that the remainder of creation has been made for him. If any opponent of the theory of plurality of worlds has advocated this idea, he has been clearly mistaken. The Sun and Moon and stars are in their respective ways useful to man,—and that usefulness is doubtless the result of design,—but to say that they were made for man, would be to state a position which is unwarranted by revelation, and, on the face of it, unjustifiable, were it for no other reason than this, that millions of the shining orbs are even beyond the sweep of telescope of strongest power. It was for God's pleasure, not man's pleasure, they are and were created, and to say otherwise is a mistake which does not flow out of the negation of plurality.

Clearing, therefore, the atmosphere of these various positions or suggestions, the question is reduced in reality to narrow limits, and it is well, before entering upon what are its real merits, we should postulate some positions which will aid us in arriving at a just determination.

And, in the first place, it will not do to say that, as all things are possible with God, so God *could* give to each and every orb on the Universe life, and life as it exists on the Earth. To attain the conclusion for such plurality would require the establishment of another premiss, which, as we have seen, cannot be predicated, that life is the proper object or utility of all such orbs

—the design for which they were all created. The Gordian knot must not be so cut.

In the second place, neither will it do to say that, even supposing the other orbs were in all respects and conditions similar to the Earth (however much such a fact would prepare our minds to receive the information), that life must exist as a necessary sequence, and for this simple reason, that life, whether of plants or animals, must be originated. Buffon, no doubt, says that wherever there is heat there must be life, and the author of the *Vestiges of Creation* would have us believe that life presses in everywhere, in pursuance of an original divine law. But these are mere assertions, they are not facts, or admitted facts, and, as we shall afterwards see, the position is disproved by the investigations which have been made.

On the other hand, in the third place, seeing that the whole doctrine of plurality is an argument of analogy, derived from the fact of life upon the Earth, so wherever we find a situation of matters which would be inimical to or destructive of life on Earth, all presumption for life must there cease. For example, we find here that life cannot exist in fire or in ice, and hence we are entitled to conclude it cannot exist anywhere else in fire or in ice—in extreme heat or in extreme cold.

In the fourth place, the demonstration that even *one* other world is uninhabitable destroys conclusion for *all* worlds being habitable, and the wider this demonstration, the more cogent will be that conclusion.

These obvious propositions being stated, lead directly

to an inquiry into the condition of the surrounding orbs. And in approaching the subject systematically, we are not only naturally led to commence with the central luminary of our system, but by doing so it will be found largely to clear the way, and to tell most forcibly upon what remains, and thus upon the whole question.

We shall combine in the next section the Sun, the Stars, and the supposed Stellar Systems.

III.

1. THE SUN.

NOTWITHSTANDING Fontenelle advocated the peopling of all the planets, he freely abandoned claim for the Sun.

'After Mercury,' he says,[1] on his fourth evening with the Marchioness, 'comes the Sun; but there is no possibility of peopling it, nor any room left for a wherefore. By the Earth, which is inhabited, we judge that other bodies of the same nature may be likewise inhabited; but the Sun is a body not like the Earth, or any of the planets; the Sun is the source or fountain of light, which, though it is sent from one planet to another, and receives several alterations by the way, yet all originally proceeds from the Sun; he draws from himself that precious substance which he emits from all sides, and which reflects when it meets with a solid body, and spreads from one planet to another those long and vast trains of light which cross, strike through, and intermingle in a thousand different fashions, and make, if I may so say, the richest tissues in the world. The Sun, likewise, is placed in the centre, from whence, with most convenience, he may equally distribute and animate by his heat; it is, then, a particular body, but what kind of body has often puzzled better heads than mine. It was thought formerly a body of pure fire, and that opinion passed current till the beginning of this age, when they perceived several spots on its surface. A little after they had discovered new planets, as we shall presently hear of, these, some said, were the spots; for those planets moving round the Sun, when

[1] *Plurality of Worlds*, ed. 1749, p. 102.

they turned their dark half to us, must necessarily hide part of it; and had not the learned with these pretended planets made their court before to most of the princes in Europe, giving the name of this prince to one, and of that prince to another planet, I believe they would have quarrelled who should be master of these spots, that they might have named them as they pleased.'

'I cannot approve that notion; it was but the other day,' says she, ' you were describing the Moon, and called several places by the names of the most famous astronomers. I was pleased with the fancy; for since the princes have seized on the Earth, it is fit the philosophers, who are as proud as the best of them, should reserve the Heavens for themselves, without any competitors.'

'Oh,' answered I, 'trouble not yourself, the philosophers make the best advantage of their territories; and if they part with the least star it is upon very good terms; but the spots on the Sun are fallen to nothing; it is now discovered that they are not planets, but clouds, streams, or dross, which rise upon the Sun, sometimes in a great quantity, sometimes in a less; sometimes they are dark, sometimes clear, sometimes they continue a great while, and sometimes they disappear as long. It seems the Sun is a liquid matter, some think of melted gold, which seems to boil over continually, and by the force of its motion casts the scum or dross on its surface, where it is consumed and others rise. Imagine, then, what strange bodies these are, when some of them are as big as the Earth; what a vast quantity must there be of this melted gold, and what must be the extent of this great sea of light and fire which they call the Sun? Others say the Sun appears, through their telescopes, full of mountains, which vomit fire continually, and are joined together like millions of Etnas. Yet there are those who say these burning mountains are pure vision, caused by a fault in the optics; but what shall we credit, if we must distrust our telescopes, to which we owe the knowledge of so many new objects? But let the Sun be what it will, it cannot be at all proper for habitation; and what pity that is, for how pleasant would it be? You might then be at the centre of the Universe, where you would see all the planets turn regularly about you; but now we are only possessed with extravagant fancies, because we do not stand in the proper place; there is but one place in the world where the study or knowledge of the stars is easily obtained, and what pity it is there is nobody there.'

'You forget yourself, sure,' says she; 'were you in the Sun you would see nothing, neither planets nor fixed stars. Does not the Sun efface all? So that, could there be any inhabitants there, they might justly think themselves the only people in nature.'

'I own my mistake, madam. I was thinking of the situation of the Sun, and not of the effect of its light. I thank you for your correction, but must take the freedom to tell you that you are in an error as well as myself; for were there inhabitants in the Sun, they would not see at all. Either they could not bear the strength of its light, or, for want of a due distance, they could not receive it; so that, things well considered, all the people there must be stone blind, which is another reason why the Sun cannot be inhabited. But let us pursue our voyage.'

While accepting Fontenelle's conclusion, we cannot help, in the light of present knowledge, smiling at some of the ideas mentioned as then entertained regarding the Sun, particularly the idea that it is composed of molten gold; the fact being that gold appears to be altogether wanting in the Sun. The passage, however, may be taken as a favourable specimen of the style of discussion characterizing the book.

Huygens took a different view.[1] He said that the Sun was hot and fiery, and that our bodies could not live in such a furnace (a fact beyond dispute), and that its inhabitants must be a new kind of animal made

[1] *Cosmothereos*, p. 143.

with a noble end and purpose,—a notion adopted by some modern writers who have thought that the heat difficulty was met by supposing the beings inhabiting the Sun are differently constituted from man. But when Wilson, in 1774, announced the opinion that the spots on the Sun were depressions on its surface, and that the black spot in the centre is the nucleus or opaque body of the Sun disclosed by this depression or void, a new theory was started, that the supposed interior nucleus, surrounded by two strata of clouds, is inhabited, and that the inner stratum is a cool barrier interposed to protect the inhabitants from the heat and brilliancy of the other stratum or outer covering constituting the region of light and heat. There was a difficulty, however, here, because the cloud which protected from the fierce rays of the Sun would exclude sight of the Heavens. But this objection was conveniently overcome by according to the inhabitants the power of peeping through the holes formed by the continually changing spots. Sir William Herschel and M. Arago both lent the weight of their authority to this astounding idea.

Sir David Brewster, discussing the subject in the fifth chapter of his *More Worlds than One*, mentions the general grounds of these opinions, and adds that—

'The probability of the Sun being inhabited is doubtless greatly increased by the simple consideration of its enormous size. Admitting, with Sir William Herschel, that the Sun may have a temperature adapted even for human constitutions, it is difficult to believe that a globe of such magnificence, 882,000 miles in diameter, upwards of 111 times the size of our Earth, and 1,384,472 times its bulk, should occupy so distinguished a place without intelligent beings to study and admire the grand arrangements which exist around them; and it would be still more difficult to believe, if it is inhabited, that a domain so extensive, so blessed with perpetual light, is not occupied by the highest order of intelligence.'[1]

Powell strikes out the view that the solar heat is of peculiar nature, and—

'Is conveyed, as it were, in the rays of light as a vehicle, and never becomes sensible as heat till the light is absorbed. It is therefore probable that these rays may owe their extrication from the Sun *to some other cause than elevation of temperature.*'[2]

In other words, this would mean that there is no such heat in the Sun itself as to forbid existence in it.

This was directly opposed to the opinion of Dr. Lardner, who said :—

'The Sun is a vast globe invested with an ocean, or rather an atmosphere of flame, in which the most astonishing convulsions and eruptions are continually manifested. Here is no moderated and regulated temperature, no alternations of light and darkness, no succession of seasons, no varieties of climate, no divisions of land and water. The Sun is, in fact, a *vast globular furnace*, the heat evolved from each square foot of which is seven times greater than the heat which issues from a square foot of the fiercest blast furnace. How utterly removed from all analogy with the Earth such a globe of fire must be, is apparent.'[3]

With this opinion Sir John Herschel apparently coincided in a work published by him in 1830; when alluding to the researches of De la Roche on the solar heat, he says :—

'This discovery is very important, as it establishes a community of nature between solar and terrestrial heat,

---

[1] *More Worlds than One*, p. 102.   [2] *Ibid.* p. 224.
[3] Quoted in Philip's *Worlds beyond the Earth*, p. 253.

while at the same time it leads us to regard the actual temperature of the Sun as far exceeding that of any earthly flame.'[1]

Dr. Dick, who disavows any decided opinion on one side or the other regarding the Sun's habitability, but who is extremely unwilling to give it up, tells us :—

'For anything we know to the contrary, or can demonstrate, the Sun may be one of the *most splendid and delightful regions of the Universe, and scenes of magnificence and grandeur may be there displayed far surpassing anything that is to be found in the planets* which revolve around it, and its population may as far exceed in number that of other worlds as the immense size of this globe exceeds that of all the other bodies in the system. But, on the other hand, we know too little of the nature and constitution of the Sun and the plans of divine wisdom to warrant us in making any positive assertions on this point.'[2]

But these speculations regarding the habitability of the Sun were all given to the world during a period of ignorance of its true nature. The discoveries which have of very recent years been made, principally by aid of the spectroscope, have put us in possession of such an amount of acquaintance with the actual constitution of the Sun, as enables us at once to see that no room can be left to doubt that the conditions prevailing in it are utterly inimical to life.

In the previous chapters on the nature and constitution of the Sun, I have entered fully upon the results of recent investigations, and it is only necessary here to state briefly what these are so far as bearing on the subject involved. They are the following :—

1. The idea of the black spots disclosing an opaque

[1] *Natural Philosophy*, p. 315. [2] *Celestial Scenery*, p. 220.

interior body is altogether exploded. These spots, varying from day to day in size and shape and position, are simply temporary depressions of the photosphere or shining cover, produced by a downpour of metallic gases, and the black central spot is nothing more than the metallic mass at the bottom of the well formed by such downpour.

2. The Sun is surrounded by layers of gases and metallic fumes or vapour to a height of at least 500,000 miles above its shining surface. These do not consist of gases capable of sustaining life. On the contrary, they are, to an enormous height, not merely partly metallic and partly hydrogen, and partly gases unknown to us here, but they are incandescent, or hot or burning flame.

3. The photosphere, or what might be termed the crust or shining cover of the Sun, is itself gaseous, and is a mass of gigantic burning waves, produced by motion of fiery currents rushing down or rushing up with a speed of thousands of miles in a second, causing continual displacement of its surface.

4. The surface is further subjected to the formation of gigantic spots or frightful chasms, causing most rapid separation and reunion of parts to an extent hardly conceivable.

5. It is also in such a state of disintegration, that apparently the portion of covering at the equator moves round with a rapidity of course differing materially from the rate of revolution nearer the poles.

6. It is also swept by tornadoes of such violence as to be comparable to nothing on Earth,—7000 miles per minute having on occasions been estimated for this frightful force.

7. The Sun shines in its own and not a borrowed light, and the power of light-rays is enormous.

8. The heat of the photosphere of the Sun has been estimated, at the lowest computation, to be many times that necessary to boil water, or even to melt iron.

Putting aside as superfluous the consideration of the effect of the tremendous pressure which the power of gravity would have upon inhabitants of the Sun at its surface,—a point which has sometimes been made also regarding Jupiter and the other larger planets,—it is therefore plainly impossible to hold that life can exist in the Sun. Animals could not live; plants, the food of animals, could not grow. Even supposing animals could possibly live in the heat, they would be continually, with tremendous vehemence, shot down or shot up through the photosphere, or blown about on its surface in one direction or another. It is evident to demonstration that life in the Sun cannot be. The circumstances which have been stated would subject the thought to a *reductio ad absurdum.*

Sir David Brewster found it difficult to believe that a globe of such dimensions could exist without being inhabited by intelligent beings. But from what has been stated we see beyond doubt, according to all that regulates animal existence, life on that fiery body is impossible; and, with much better cause, we can use the words of Huygens, and say: 'We have gained a point.' We not only depopulate the Sun, but in doing so we depopulate a body which is at least 600 times larger than all the other bodies of the solar system put together. Not only so, but we thus deprive a star (for the Sun is a true star) of life. The importance of this conclusion becomes manifest when we proceed to consider, in the next place—

2. THE FIXED STARS.

BEFORE the discoveries made by means of the spectroscope, it was a fact admitted on all hands, that the fixed stars must of necessity be self-luminous. From the immense distances at which they were situated no reflected light could have reached us from them. Even the advocates of plurality were therefore constrained to admit that this circumstance was adverse to their habitability. Thus Powell stated—

'That the stars are self-luminous, and therefore (it might be argued) probably consist of matter in a state of incandescence or combustion of some kind, is the strongest fact against their being inhabited. This, however, is no more than the same argument relative to the Sun, and, like it, susceptible of the same answer, from conjecture, as to the existence of an exterior photosphere.'[1]

In other words, some of the advocates of plurality believed that the fixed stars might be, as they imagined the Sun is, composed of an opaque solid nucleus capable of life, surrounded by a shining envelope or photosphere. But they generally, if not universally, adopted another or further opinion, to which I shall immediately advert. In the meantime it is necessary, by a few words, to dispose of this view.

Dr. Huggins, in his lecture on 'Spectrum Analysis in its application to the Heavenly Bodies,' delivered in Manchester in November 1870, related the means taken by

[1] *Unity of Worlds*, p. 213.

him and the late Dr. William Allan Miller to analyse the heavenly bodies by application of the spectroscope, aided by the telescope. He then showed the spectrum of the stars Aldebaran and Betelgeux, pointed out what substances the spectra disclosed, and stated[1] that we thus learnt that they were subjected to the same force as in the solar system,—the force of heat,—and that they emitted light in the same way as terrestrial substances emit light. They had examined some fifty of the stars, and they were found to differ from one another, but were all constructed upon the same type,—all of them contained some terrestrial substances, but apparently in different proportions, and also, it might be, many other new elements entirely unknown to us. They subsequently, I believe, extended their investigations to a much larger number of stars.

From another source we learn that—

'While Huggins and Miller had thus been investigating about a hundred of the brightest stars, Secchi, favoured above his English fellow-labourers by the purity of an Italian sky, had already extended his observations over more than 500 fixed stars, and gave the results to the world, in 1867, in his work, entitled *Catalogo delle Stelle di cui si è determinato lo spettro luminoso, all' osservatorio del Collegio Romano.* Since then above a hundred more stars have been added to this catalogue by this industrious astronomer, so that there exists at present a rich mass of spectrum observations of the fixed stars, which Secchi has so far provisionally arranged as to be able to group them into four principal types, into which all stars, with only a very few remarkable exceptions, may be classified.'[2]

Those who have thus examined the stars seem to be agreed that they are in constitution similar to the Sun.[3]

---

[1] Pp. 39, 40.   [2] Schellen's *Spectrum Analysis*, p. 495.
[3] We have said before that the Sun is a star; and this statement may

Like the Sun, they possess an incandescent photosphere, and are surrounded by a hot or incandescent vaporous atmosphere. But the substances of which their respective atmospheres are composed differ in each. In other words, while the stars all shine in consequence of their great heat and burning atmosphere, the vapour which manifests them springs from various substances, some of them known to us on Earth, and some of them unknown. Dr. Schellen, whose work on *Spectrum Analysis* (1872) is one of the most recent, and the translation of which has been edited by Dr. Huggins, says :—

'As the spectra of the stars bear, in general, a marked resemblance to the spectrum of the Sun, being continuous and crossed by dark lines, there is every reason for applying Kirchhoff's theory also to the fixed stars, and for accepting the same explanation of these similar phenomena that we have already accepted for the Sun. By the supposition that the vaporous *incandescent* photosphere of a star contains, or is surrounded by, heated vapours, which absorb the same rays of light which they would emit when self-luminous, we may discover from the dark lines in the stellar spectra the substances which are contained in the photosphere or atmosphere of each star.'[1]

now be reversed, and we may say that the stars are distant suns. Their spectra are similar to that of the Sun; but the greater difficulty of examining them renders the results comparatively meagre. The following may be cited as examples of stars which have been examined spectroscopically :—

Aldebaran contains hydrogen, sodium, magnesium, calcium, iron, bismuth, tellurium, antimony, mercury.

Betelguex (α Orionis) contains sodium, magnesium, calcium, iron, bismuth, thallium (?).

Seruis contains sodium, magnesium, iron, hydrogen.[2]

---

[1] Schellen, p. 489.
[2] Skertchley's *Physical System of Universe*, p. 79.

Then, after stating the results of investigation, he concludes thus:—

'From all these observations, it may be concluded that at least the brightest stars have a physical constitution similar to that of our Sun. Their light radiates, like that of the Sun, from matter in a state of intense incandescence, and passes in like manner through an atmosphere of absorptive vapours. Notwithstanding this general conformity of structure, there is yet a great difference in the constitution of individual stars; the grouping of the various elements is peculiar and characteristic for each star; and we must suppose that even these individual peculiarities are in necessary accordance with the special object of the star's existence, and its adaptation to the animal life of the planetary worlds by which it is surrounded.'[1]

The last observation has no necessary connection with the present branch of inquiry, and it assumes both the existence of planetary worlds and life upon them.

Roscoe, in his *Lectures on Spectrum Analysis*, even more clearly expresses himself thus:—

'We have then now arrived at a distinct understanding of the physical constitution of the fixed stars; they consist of a white hot nucleus, giving off a continuous spectrum, surrounded by an incandescent atmosphere, in which exist the absorbent vapours of the particular metals. These results are interesting, as bearing on Laplace's nebular theory, because they show that the visible Universe is mainly composed of the same elementary constituents, although certain of the stars differ from one another widely in their chemical constitution.'[2]

Now we have seen that the Sun exists in conditions

[1] P. 506. [2] Roscoe's *Lectures on Spectrum Analysis*, p. 237.

which render life in it impossible. Of course the accumulation of such conditions renders it all the more impossible; but it may be said each different condition is of itself sufficient to prevent even presumption for life. The stars may not be exactly alike to the Sun in all these respects; but the simple fact of the intense heat contained in them is sufficient to satisfy us of the impossibility of life, either animal or vegetable (the support of animal) life, existing on any of them.

Here, then, is a second point established. Every one of the stars which we see, and every one of the stars which the telescope brings to sight, is uninhabitable. This inevitable conclusion sweeps the firmament, saving the planets of our system, which fall to be separately and carefully considered in detail, and excepting the comets and the nebulæ, both of which, the spectroscope reveals, are gaseous bodies.

'The nebulæ,' says Skertchley,[1] 'are found to consist of incandescent gases, chiefly hydrogen, and some of the comets of a compound of carbon.'

Plainly neither comets nor nebulæ are habitable, and any fanciful supposition that they, or at least the first mentioned, are, which formerly may have been held, has been now, I think, fairly abandoned; but afterwards we shall note a few facts regarding comets, which will show how entirely their condition is irreconcilable with the idea of their being the abodes of life.

Yet the fact, now so indisputable, that the stars themselves are uninhabitable, only removes the question to another stage. Both before the application of the spec-

[1] Skertchley's *Physical System*, p. 79.

troscope to the stars and since, the view has been taken that every star is, like the Sun, attended by a system of planets, so that the star exists for the sake of its planets, and these hypothetical planets exist for the sake of the life with which they are presumed to be endowed. The next point for consideration therefore is—

3. THE SUPPOSED STELLAR PLANETARY SYSTEMS.

FERGUSON, in his *Astronomy*,[1] after showing 'that the stars are of the same nature with the Sun,' proceeds to say :—

'It is noways probable that the Almighty, who always acts with infinite wisdom, and does nothing in vain, should create so many glorious Suns, fit for so many important purposes, and place them at such distances from one another, without proper objects near enough to be benefited by their influences. Whoever imagines they were created only to give a faint glimmering light to the inhabitants of this globe, must have a very superficial knowledge of astronomy, and a mean opinion of the divine wisdom, since by an infinitely less exertion of creating power the Deity could have given our Earth much more light by one single additional moon. Instead, then, of one Sun and one World only in the Universe, as the unskilful in astronomy imagine, that science discovers to us such an inconceivable number of suns, systems, and worlds, dispersed through boundless space, that if our Sun, with all the planets, moons, and comets belonging to it, were annihilated they would be no more missed by an eye that could take in the whole creation than a grain of sand from the sea-shore.'

Sir David Brewster,[2] a century later, expressed himself more dogmatically thus :—

'We are compelled to draw the conclusion that wherever there is a Sun, a gigantic sphere, shining by its own light, and either fixed or moveable in space, there *must be a planetary system, and wherever there is a planetary system there must be life and intelligence.*'

[1] 6th edition, 1778, p. 3. [2] *More Worlds than One*, p. 170.

These are peremptory statements which are inadmissible. Besides assuming that the object of creating worlds is for life, they also assume what is erroneous, that a Sun must necessarily be attended with planets. There is indeed a possibility, especially having regard to the claims of the nebular hypothesis, that at least many of the stars are surrounded by planets. But this is not to be taken for granted. For, in the first place, we have no knowledge that they in point of fact are so. Whether we shall ever be able to ascertain the fact by means of the inventions and discoveries of the future we cannot tell, and the fact, therefore, in our present state of knowledge, is not to be assumed; but this much may be said, that in all likelihood, as there are nebulæ —sheets of nebular or star-forming material—which remain unconverted into stars, and are nothing but huge tracts of hot gas, so there are many stars which have not reached the stage of throwing off planets, while perhaps there are many others which never will. According to the nebular hypothesis, which, as applicable to our system, seems to be placed beyond doubt, there was a time when our Sun existed without a single planet circling round him; and in this condition is it not possible that, for example, such a huge sun as Sirius may now be? It is a circumstance far from improbable. On the other hand, we know that there are great orbs in space which are dark, and therefore cannot be seen by us; and knowing, as we do, that our Sun is gradually cooling, and will some day, however distant, become black, we may suppose that there are orbs no longer suns which are, if any are, attended by planets, and yet nothing can be more certain, in the view of all the facts connected with the maintenance of life, than that such planets are uninhabitable. Therefore Sir

David Brewster's imperative assertions are clearly wrong.

But there is a further fact which has been much discussed as bearing on the impossibility that every star can be attended by planets. It is that of double or binary stars—stars which are connected by circling round each other, or round a common centre in regular orbits. Professsor Nichol thus explains the phenomenon:—

'Have you ever,' he says, 'walked in a mood of tranquil thought along the side of a quiet river, whose waving banks reflect a thousand currents by the intermingling of which numerous dimples or whirlpools are produced, their easy glide only waking the river's stillness? Have you seen these dimples follow and pursue each other as if in gambol, or watched the phenomenon of the near approach of two or three? Then have you witnessed the secret of the mystery of the double and triple stars.'[1]

The subject of such stars (also found triple, quadruple, and multiple) is fully discussed by Sir John Herschel in his *Outlines of Astronomy*, § 833 et seq. But the point which is of main importance in this inquiry is the fact of the short periods, in general, of their common revolution. Sir John has given[2] a table of the chief results of investigation in regard to binary stars, in one column of which the calculated duration of their revolutions is stated, and it appears that the majority are under 100 years, one being so low, according to Madler's estimate, as only 30 years.[3] The highest reaches up to

---

[1] Nichol's *Architecture of Heavens*, p. 172.
[2] *Outlines*, § 843, p. 615.
[3] Sirius would appear, from recent investigations, to be a binary star, its nearly dark companion being distant from it rather more than the

736 years. It would seem, however, that other double stars have longer periods. Now, bearing in mind that Saturn revolves round the Sun in about 30 years, and Neptune in about 165 years, and if a planet outside Neptune be, as it is expected, discovered, its revolution will probably extend to about 320 years, it seems manifest that some, if not many, of these binary stars are nearer to each other than some of our planets are to the Sun, and none of them are removed from each other to distances many times greater than the most remote planet is from the Sun. And Sir John Herschel makes this important observation with reference to the planets supposed to revolve round these binary suns:—

'Unless closely nestled under the protecting wing of their immediate superior, the sweep of their other sun in its perihelion passage round their own might carry them off, or whirl them into orbits utterly incompatible with the conditions necessary for the existence of their inhabitants.'[1]

Dr. Whewell in his essay on *Plurality* follows up this remark by observing—

'A system of planets revolving around or among a pair of suns which are at the same time revolving about one another is so complex a scheme, so impossible to arrange in a stable manner, that the assumption of the existence of such schemes without a vestige of evidence can hardly require confutation.'[2]

And then referring to the 'sweep' of one upon the supposed planets of the other, he says that the only safety would be in supposing them placed so close to

distance of Neptune from the Sun. Helemholz, *Popular Lectures*, 2d series, p. 151.

[1] *Outlines*, § 847, p. 618.   [2] *Plurality*, chap. viii. p. 259.

one Sun as to be out of reach of the other, and this might involve their plunging through a luminous vapour and being arrested, impeded, attracted, and lost in the central nucleus. Equally fatal, at least to inhabitants, would, I think, be the close proximity to such suns.

Sir David Brewster's method of overcoming the objection is curious. He concludes from analogy that—

'If Madler's speculation is correct, our Sun and the star Alcyone form a binary system, and therefore, since our Sun is attended with planets, and one of these inhabited, we are entitled by analogy to conclude that all other binary systems have planets, at least round one of their suns, and that one of these planets is the seat of vegetable and animal life.'[1]

Sir David, however, forgets that even were our Sun a binary star, its revolution takes not 30, nor even 30,000 years, but 18 millions of years; and even the star nearest to us is immensely removed from the power of exercising attraction over the planets of the solar system, so that there can be no reasoning by analogy from our Sun to any of the binary stars.

But the matter does not rest here. For Sir John Herschel has, by taking the discovered parallax of the double star 61 Cygni, and calculating the distance between the two stars, been enabled 'to calculate the sum of the masses of the two stars, which on these data we find to be 0·353, the mass of our Sun being 1.' Thus the mass of each of the two stars is somewhat about one-sixth of the mass of the Sun, and I think it is only a reasonable conclusion, that being diffused so largely as to appear a star to us at its great distance, the star is six times less dense; in other words, it is

[1] *More Worlds than One*, chap. x. p. 169.

six times more vapourous than the Sun, which has only the one-fourth of the density of the Earth, and hence it is in a condition in which the Sun is not at present, whatever it may have been when he first threw off the outermost planet. Thus, it may be inferred, that this double star has not yet reached, if it ever will reach, the condition of capacity for forming planets. Taking this view, the probabilities are that there are many in similar condition, and we thus arrive at a conclusion adverse to these stars having planets, not to say habitable planets, moving round them, and therefore were not made to sustain life by means of a surrounding planetary system.

Now this is important, not merely as respects these two individual stars making up a binary star, but as bearing upon the whole question of starry systems. If it be satisfactorily shown that some stars cannot presumably be supposed to be surrounded by planets, then a conclusion cannot be arrived at that all stars have a planetary system, and this assists to some extent, in the absence of every evidence to the contrary, to destroy the presumption for any star being so attended.

But it will be said that if these stars are not now surrounded by planets, they will be—they are in a state of preparation. It may be so. We cannot tell. One very awkward preparation for the double stars would, however, seem to be that (in Professor Nichol's words) 'they must, on approximation, act upon each other as two wheels, so that a revolution of each around the other must immediately supervene and increase in rapidity until, by external pressure, they are *forced into one*.' So that before they can begin to acquire the

ARE THEIR PLANETS HABITABLE? 163

condition in which they may be enabled to throw off double planetary rings, they must come into a collision of which it would be terrible to think. Perhaps such a collision would be fatal to the formation of planets. But if otherwise, all that can be said is, that at some future time they may become attended by planets; meantime, they are not, and meantime these stars, which are numerous, do not support life.

In any view, then, what has been said goes to this, that at present some, or it may be many, stars *are* without planets, whether habitable or not. However, without making any positive affirmation on a point so far beyond our discovery, I have said sufficient to discredit the existence of stellar planetary systems, and for holding that in any view they cannot be universal, and that we can only deal with the subject hypothetically for the purposes and further progress of the argument.

But now arises the question : assuming, for the sake of argument, that the stars, or many of the stars, are attended by planets, Can these planets be presumed habitable ? The only mode of answering or considering this question is by taking the analogy of the solar system, and seeing how far, if at all, it will bear out the theory of plurality. The subject, then, now for consideration is—

Are the supposed stellar planets by analogy to be presumed life bearers, and inhabited by rational beings akin to man ?

Here we are met at the very threshold of inquiry by a very staggering fact.

For the researches of the spectroscopists have revealed

to us what might only previously have been matter of supposition, that the stars while similar, or so far similar, in constitution, and possessing many properties or substances in common, differ not only from each other, but from our own Sun, in the gases of which they are respectively composed. This appears from the quotation I have already made from Schellen's *Spectrum Analysis*, and I would refer to the subject as treated by him at p. 489 et seq., from which it will be seen that, in accordance with Father Secchi's arrangements, he groups the stars into four principal types, besides others of apparently rarer occurrence; and he proceeds to mention the coloured stars,—stars which are of all shades of red, green, blue, and violet, and are most conspicuous among the double stars, 'the number of which already (1872) exceeds 6000,' and with the phenomena of variable and new or temporary stars.

Professor Roscoe has also dealt with the subject;[1] and when referring to the investigations of Dr. Huggins and Dr. Miller into the spectra of Aldebaran and Betelgeux,[2] the star known as *a* in the constellation of Orion, makes these observations:—

'In the first place, then, the result at which we have arrived is, that the constitution of the star light, although not identical with the light given off by the Sun, is yet similar; that is to say, the light of a fixed star gives off a continuous spectrum, interspersed by dark shadows or bands; and hence the conclusion we come to is, that the physical constitution of the fixed stars is similar to that of our Sun, that their light also emanates from intensely white hot matter, and passes through an atmosphere of absorbent vapours — in fact, that the stars are suns of different systems. We find, for instance, in these two particular stars to which I am now referring, the D line

[1] *Lectures on Spectrum Analysis*, p. 231 et seq.   [2] *Antea*, p. 152.

caused by sodium exists; the three lines which we know as *b* are produced by luminous vapour of magnesium. The lines of these substances exactly agree in position with the dark stellar lines, hence both sodium and magnesium are present in the atmosphere of these far distant stars. We also find in Aldebaran that two hydrogen lines, C and F, are present, but if you look at the spectrum of α Orionis, we find that the hydrogen lines C and F are wanting. Hence we come to the conclusion that hydrogen is present in the atmosphere of the Sun and in that of Aldebaran, but that it is wanting in that of Betelgeux. And so I might show you that silver is not present in Aldebaran, nor seen in α Orionis, but that four bright lines of calcium, also seen in the Sun's spectrum, are present in both stars. The lines observed in these two stars are, at least, seventy in number, and Mr. Huggins and Dr. Miller have found that in Aldebaran we have evidence of the presence of no less than nine elements—namely, (1) hydrogen, giving the lines C and F; (2) the metal sodium, giving the double line D; (3) magnesium, giving the lines *b*; (4) calcium, giving four lines; (5) iron, giving four lines and E; (6) bismuth, giving four lines (bismuth is not found in the Sun); (7) tellurium, four lines; (8) antimony is also found, three lines; and (9) mercury, four lines. Thus the element tellurium, whose name implies a purely earthly origin, is found in the star, although it does not exist in the Sun, and is very rare in this Earth. There are only two stars—Betelgeux, to which I have just referred, and another star called β Pegasi—in which the hydrogen lines are wanting; all the other stars contain hydrogen.'[1]

We are thus shown that the stars differ from each other in the elements contained in them, or in the gases into which they are by their heat thrown. While it is extremely likely that even where the elements are the same, the quantities of each element may vary very much, and this warrants certain deductions. For—

[1] Roscoe, p. 236.

1. It is not at all improbable that a very large number of the stars may not contain the ingredients, or sufficient quantities of particular ingredients, necessary to constitute a solid planet similar to the Earth. We have example in comets of planetary bodies in our solar system which the spectroscope tells us are wholly or principally gaseous,[1] are composed partly of matter gaseous and self-luminous, and partly of minute particles held together by mutual attraction and gravitation, but so rare that stars are visible through the body and tail of the comet.[2] Now, without going the length of supposing planets attached to other stars to be of a composition of such extreme tenuity as that of comets, the existence of comets shows—and the facts I shall have afterwards to state regarding certain of the members of our system confirm the thought—that it is quite possible that if stellar planets exist there may be many of them possessed of little solidity,—perhaps none of them have sufficient density to be life-supporting,—this being all the more probable in the case of stars which are themselves more gaseous, or rather, possessed of less density than the Sun.

2. Supposing, on the other hand, that other stars contain elements which are calculated to give solidity to their planetary attendants, it may be that the solidity is of such a character as would be fatal to the thought of vegetation or life. We have example in meteoric stones, which are just minute planets, of bodies of hard metallic composition which are incapable of supporting vegetation.[3]

---

[1] Roscoe, pp. 251 et seq., 297 et seq. *Manchester Science Lectures*, 1871, Huggins, pp. 44–46.

[2] Schellen, pp. 568 et seq., 579.

[3] The chief constituent of meteors which have fallen to the earth in

It is not improbable, therefore, that many others of the supposed stellar planets are equally hard, and are possessed of no aqueous vapour or water, or other gas capable of pulverizing or of reducing them to a condition susceptible of promoting vegetable growth, and consequently they would be incapable of sustaining animal life.

3. The elements in others of the stars may be such as will not produce an atmosphere adapted for life. Here again the comets come to our aid. For Dr. Huggins' examination of a comet by the spectroscope [1] elicited the spectrum of carbon—and carbon volatilized, it is presumed, is destructive to animal life. But we are not left to form an opinion from comets, for, as we shall afterwards see, we have an example in one at least, if not more, of the planets of our solar system of the presence of a deadly atmosphere. Meantime it is important to bear in mind that as it is not every soil or substance in which plants will grow, so it is not every atmosphere which plants and animals can breathe. And it is the more necessary to say so because it seems to have been taken by many for granted that if an atmosphere of any kind be found surrounding a planet, the planet must forthwith be supposed capable of maintaining life. The

the form of meteoric stones is 'metallic iron, mixed with various silicious compounds; in combination with iron, nickel is always found, and sometimes also cobalt, copper, tin, and chromium among the sillicates. Olivine is specially worthy of remark as a mineral very abundant in volcanic rocks, as also augite. There have also been found in the meteoric stones hitherto examined oxygen, hydrogen, sulphur, phosphorus, carbon, aluminium, magnesium, calcium, sodium, potassium, manganese, titanium, lead, lithium, and strontium.' Schellen, pp. 583, 584. See also Phipson's *Mysteries of Nature*, p. 239 et seq., and Humboldt's *Cosmos*, i. p. 117 et seq., iv. p. 592.

[1] Roscoe, *Lectures on Spectrum Analysis*, p. 300.

atmosphere which we find capable of maintaining life consists of oxygen, nitrogen, and carbonic acid in certain given proportions, varying in a very slight degree in different situations, together with aqueous vapour and other ingredients, to a minute and varying extent, hardly to be reckoned, and generally ignored.[1] No animal can live in nitrogen; and although oxygen nourishes life, too much of it, or too long inhaling of it alone will, by the production of an excess of vital action, produce death. As Cooley says:—

'In an atmosphere of pure oxygen gas vital action would go on so fast as to terminate in an instant, destroyed by its excessive energy.'[2]

An admixture of nitrogen is therefore needful to reduce the stimulus to a proper and healthful degree. In like manner, the minute proportion of carbonic acid must not be exceeded. A writer on colliery management says:—

'Let it be increased to one hundredth, and its effects begin to be felt in headache, langour, and general depression. In the case of a recent accident in England, one of the Government Inspectors stated that 2 per cent. of carbonic acid was most injurious, and 4 per cent. would speedily kill.'[3]

---

[1] There is minute difference of statement of the proportions in different books. In ten or twelve different books as many differing statements will be given, but the discrepancies are not in the least material to the present question. Commonly the proportions stated are nitrogen 79, oxygen 21 in 100; but a half per cent. is allowed by Herschel for carbonic acid ·05, aqueous vapour ·45.—Herschel's *Meteorology*, p. 17. The amount of carbonic acid is by different writers variously estimated. See Hartley's *Air*, pp. 29, 30; Hopkins' *Atmospheric Changes*, p. 1; Griffith's *Chemistry of Fire, Air*, etc., p. 186; Ansted's *Physical Geography*, p. 233, and many others.

[2] Cooley, *Physical Geography*, p. 108.

[3] *Colliery Management*, by Jonathan Hyslop, p. 391.

In like manner it may be stated[1] that nitric oxide gas is composed of 55·95 oxygen, 44·05 nitrogen, and animals are instantly suffocated on being immersed in it. Further, according to the experiments of Lavoisier and Seguin, a mixture containing 40 of azotic, 45 oxygen, and 15 carbonic acid destroyed animals confined in it.[2]

Such being the ingredients of life-supporting air, and such the delicacy of their proportional admixture that a deviation from it is fatal, it is easy to see that even where conditions in stars are or may be supposed to be somewhat similar to those of the Sun, and there has been produced on planets belonging to their systems an atmosphere, it by no means follows that such atmosphere is capable of supporting life, a very little variation would be effectual to destroy, or would fail to sustain it. Now it is hardly possible to suppose two suns exactly alike, or two systems alike, or like our own. For there are many circumstances which will combine to produce distinctive variety of condition, and these distinctive varieties will necessarily affect the nature of the atmosphere, if any, which will be produced in the planets. Among them will mainly be the shade of difference in the composition of the individual stars, both of kind and amount,—the differences of heat and light and power of the stellar rays, the distances of the several planets and their different compositions, and the various other influences to which they are subjected. In many cases there is also the difference of colour of the stars, and even the combination of colours in binary systems.

So that we can well suppose the chances for an atmosphere similar to our own are exceedingly slight. Probably, upon the theory of probabilities, the chance might be calculated. In many cases, too, the atmo-

---

[1] Robertson's *Atmosphere*, ii. p. 42. [2] *Ibid.* ii. p. 87.

sphere may be wholly of a positively noxious gas, such as carbonic acid.

We have thus seen—
1. That the Sun cannot be inhabited.
2. That neither can the stars be.
3. That it is not to be assumed that the stars are surrounded by planetary systems, and that in the case of binary stars the probabilities are against it, and by displacing such systems from some stars a presumption arises against any.
4. Separately, that assuming the possibility of other planetary systems, many of the stars may not have reached the state of throwing off planets, while others may have reached the planet-bearing stage, and passed to that stage when their light has burnt out, so that these planets cannot be life supporting.
5. Supposing other planetary systems to exist, the revelations of the spectroscope are adverse to the idea of the planets of such systems being habitable, in respect—

(1) That many stars may not be capable of producing *solid* planets.

(2) That where capable of producing solid planets the solidity may be of a character unfitted to support life, either vegetable or animal.

(3) That the stars, or many of them, may be incapable of producing an atmosphere in their planets which is life supporting, and this incapacity is to be presumed generally throughout them all, in respect of the difference of composition, situation, and other conditions subsisting between the stars and our Sun.

IV.

WE are now brought more nearly to consider what is really the main question. As the habitability of the stars, or rather of their supposed planetary systems, is inferred, and only inferred from analogy to the Earth and the members of the solar system, we have to ask, primarily, What are the facts attendant upon Life on the Earth ? and then, What is the condition of each of the other planets ? Do circumstances warrant conclusion for their being habitable ? and from their presumable condition, what are we to infer regarding stellar planets ?

THE EARTH.

And in order to consider this pretty wide subject satisfactorily, we must begin by first of all looking a little at the case or predicament of the Earth on which we dwell, which has been tenanted by *rational* beings for a limited period of time.

For it is with the conception of *rational* life existing in other worlds we have truly to deal. Mr. Proctor has very justly observed :—

' It would be difficult to show that mere life without the power which man possesses of appreciating the wonders of the Universe, is a more fitting final purpose in creation than the existence of lifeless but moving masses like the suns and their attendant planets. The insect or the fish, the bird or the mammal, the minutest microscopic animalcule, or the mightiest cetacean, may afford suggestive indications of what we describe as beneficent contrivance; yet it is hard to see in what essential respect a Universe of worlds beyond our own, inhabited only by such animals, would accord better with those ideas which the believers in the plurality of worlds entertain respecting the purpose of the Almighty than a Universe with none but vegetable life, or a Universe with no life at all, yet replete with wonderful and wonderfully moving masses of matter. It is rational life alone to which the arguments of our Brewsters and Chalmers really relate."[1]

Now the first observation which falls to be made in regard to the Earth is, that man, the only rational being inhabiting it, has been its occupant for only a comparatively brief term.

Admittedly, on all hands, he was the last being placed upon the Earth. But some have claimed for him a high, a marvellous antiquity; 100,000, 300,000, 500,000, and even 9,000,000 years are among the periods which learned and sanguine men have assigned for the past duration of the human race, arriving at such startling enunciations only by means of conjecture based upon facts, or supposed facts, which in themselves afford no such data for computation. The antiquity theory seems to be specially favoured by the Evolutionists, who find it impossible to reconcile a recent date for the creation of man with the demands of Darwinism.[2]

The facts upon which those contending for antiquity have proceeded, have mainly been the discovery of

---

[1] *Our Place among Infinities*, p. 62.
[2] 'In the meanwhile, if any form of the doctrine of progressive development is correct, we must extend by long epochs the most liberal estimate that has yet been made of the antiquity of man.'—Huxley's *Man's Place in Nature*, p. 159.

human implements, or what are thought to be so, or remains of a supposed age, in conjunction with the remains of animals believed by them to have become extinct at remote times, or underlying certain layers imagined to have taken enormous periods for their deposition or formation; or, in other situations or circumstances, regarded as indicating a vast antiquity. They will be found set forth in such books as Sir Charles Lyell's somewhat cautious work on the *Antiquity of Man*. Principal Dawson of M'Gill's University, Montreal (who has examined the subject at considerable length in his *Origin of the World* (12th to 15th chapters), and satisfactorily disposes of these notions), thus summarizes the arguments from geology in favour of a great antiquity for man:—

'(1) Human remains are found in caverns under very thick stalagmite crusts, and in deposits of earth which must have accumulated before these stalagmites began to form, and when the caverns were differently situated with reference to the local drainages. (2) Remains of man are found under peat bogs which have grown so little in modern times that their antiquity, on the whole, must be very great. (3) Implements, presumably made by men, are found in river gravels so high above existing river beds, that great physical changes must have occurred since they were accumulated. (4) One case is on record where a human bone is believed to have been found under a deposit of glacial age. (5) Human remains have been found under circumstances which indicate that very important changes of level have taken place since their accumulation. (6) Human remains have been found under circumstances which indicate great changes of climate as intervening between their date and that of the modern period. (7) Man is known to have existed, in Europe at least, at the same time with some quadrupeds formerly supposed to have been extinct before his introduction. (8) The implements, weapons, etc., found in the oldest of these repositories are different from those known to have been used in historic times.'[1]

The case of the antiquaries is shortly but more fully stated by one of themselves, Mr. Thomas. Dunman, in a chapter of *Talks about Science*,[2] published since his death. In this (his lecture on ' Prehistoric Man ') he has adduced within small compass all the facts on which reliance is placed in support of great antiquity.

Many of these facts, with the inferences built upon them, have from time to time been shown to be altogether unreliable. Principal Dawson, in the work above referred to, as well as in his *Story of Earth and Man*, and his *Fossil Men and their Modern Representatives*, has dealt with them. But it has been reserved for another American, Dr. James C. Southall, in works evidencing great research and erudition, systematically to demonstrate in a manner, and with a force of reasoning which appears conclusive, that every one of the circumstances which have been relied upon is to be put aside. The first of these, a large volume of 600 closely printed pages, called *The Recent Origin of Man*, published in 1875, is a very storehouse of learning and research, and to those who desire to study the subject fully, it must afford the most ample information. But it was published in America, and is rarely seen here.[3] Probably the author had thought it too bulky for popular reading, and in 1878 published in England another, *The Epoch of the Mammoth*, which apparently is, to a great extent, an epitome of the other, and is, perhaps,

[1] *Origin of World*, 2d ed. p. 295.
[2] Dunman's *Talks about Science*, p. 36, Lecture iii.
[3] I could only find it in one library in Edinburgh—the Antiquarian.

better adapted for those who do not wish to go profoundly into the subject. At all events, it affords sufficient material for such an incidental reference to it as only is needful here. It may therefore be referred to for the amount and weight of evidence brought forward in support of the triumphant stand this American gentleman has made against the doctrines which have been in some quarters so confidently maintained. I will only mention a few of his demonstrations.

Thus he has shown that those animals supposed to have been extinct at immensely remote epochs, have all lived down to a recent or historical period; that the stalagmite, supposed to take thousands or millions of years to produce a few feet, forms, in favourable circumstances, with rapidity;[1] that the river gravel and the peat mosses, the lake dwellings and the kitchen middens, are all comparatively recent; that there was no real separation into stone, bronze, and iron ages, and, indeed, stone sometimes succeeded ages of higher civilisation, but that the use of all was intermingled or coeval partly so, and there is no vast antiquity for the stone ages.

Passing by the hoaxing to which antiquaries have on this question been subjected, as in the case of the bones of an African elephant, which for half a century were palmed off as the skeleton of a genuine Columbian Mastodon, a deception only revealed at last by a death-bed confession;[2] it may be mentioned that sometimes

[1] It may be stated, but it is by no means necessary to the question, that 'the fossils of the older Post-Pliocene must often, by the process of sorting by water, have been mixed with those of the river,' a circumstance which doubtless may explain an occasional difficulty.—Dawson's *Story*, p. 288.

[2] Paragraph in *Daily Telegraph*.

the discovery of supposed evidence of antiquity is pitilessly upset by the further discovery, at one time in juxtaposition, at another at a lower level, of a Roman coin. On one occasion a supposed ancient carving was determined, upon better examination, to be the gunwale of a modern boat. Indeed, the antiquaries who glory over what they consider to be precious 'finds,' must sometimes share the mortification of Jonathan Oldbuck as he smarted under the matter of fact explanations of Edie Ochiltree. Mr. Pengelly's beaming 'delight' over the relics of Kent's Hole, as affording proof of a hoary human antiquity, far past the conceptions of that ancient and incredible historian Moses, must, one would suppose, have been considerably dashed by the awkward revelations and observations of Drs. Southall and Dawson.

But the subject is a very wide one, and can only be thus incidentally referred to. One or two further circumstances show that man is not only of recent origin, but that he has not lived upon Earth for more than about 6000 to 7000 years.

1. Although a fossil state does by no means imply antiquity (and the dripping well of Knaresborough, Yorkshire, affords a curious example of the extreme rapidity with which things may be fossilized), there is apparently no evidence of man being found in that condition. Baron Cuvier, writing in 1812, states as matter of fact, 'It is quite undeniable that no human remains have been hitherto discovered among the extraneous fossils,'[1] and I have not observed that any human bones truly fossil, or of great antiquity, have been discovered since. What have been called the fossil

[1] Cuvier's *Theory of the Earth*, Jamieson's translation, p. 127.

skeletons of Florida, of Guadeloupe, of New Orleans, of Mentone, are either not fossil, or can be shown to belong to merely an early period of man's history, sometimes only to a few centuries back.[1]

2. Dr. Southall shows that man appeared on Earth not earlier than the later post-pliocene period; and by various converging facts and corresponding calculations made by different men, that the glacial period terminated just about 6000 years ago.[2]

3. Man must have left his Creator's hands perfect, as

[1] See Southall and Dawson, also a little book by the Rev. J. Brodie on *The Antiquity and Nature of Man.* 1864.

[2] Lieut.-Col. Drayson, who has devoted special attention to the subject of the glacial epoch, does not seem greatly to differ. 'The period spoken of in geology as the glacial period would have occurred between the dates about 6000 B.C. and about 22,000 B.C. At the date midway between these two periods would have been the height of the cold or glacial period. Hence from about 12,700 B.C. down to the present time we have the interval during which the alluvium and surface soil formations took place, a space of time which probably some geologists may consider too brief, unless they bear in mind that far more powerful causes were then at work than there are at present.'—*The Last Glacial Epoch in Geology*, by Lieut.-Col. Drayson, p. 167. At another place (p. 159), where he has placed the height of the last glacial epoch at 13,700 B.C., he tells us there would be an epoch of 8000 years subsequently, 'during which masses of frozen snow and ice would be annually floating over the Earth's surface,' bringing down the termination of the glacial period to 5700 years B.C. In dealing with the antiquity of man, the writer says: 'We may bring the mammoth period nearer our own time than those hundreds of thousands of years which have been assumed by some writers as absolutely necessary.' Mr. Prestwick has also, from geological investigations, come to a similar conclusion, as shown in the following extract 'Whilst abstaining from any general hypothesis in explanation of the phenomena, there is, however, one point to which I must refer before concluding, although I cannot at present venture beyond a few generalities respecting it. It might be supposed that in assigning to man an appearance at such a period, it would of necessity imply his existence during long ages beyond all exact calculations, for we have been apt to

perfect for his place as the leaf upon the tree, as perfect as any even of the very earliest and humblest of the animate creation was for the lowly purpose for which it was designed. His knowledge from experience would necessarily be at first nil, and for a time scanty, but his bodily frame would be faultless and finely formed—sound, strong, robustly healthy; his mental faculties correspondingly complete—vigorous, clear and penetrating, nay, doubtless, to compensate deficiency of knowledge, endowed and quickened with an instinctive sagacity to which we have now no approach. The notion that the first man was savage is a strange theory, and receives

place even the latest of our geological changes at a remote and to us unknown distance. The reasons on which such a view has been held have been mainly the great lapse of time considered requisite for the dying out of so many species of great mammals; the circumstance that many of the smaller valleys have been excavated since they lived; the presumed non-existence of man himself; and the great extent of the later and more modern accumulations. But we have in this part of Europe no succession of strata to record a gradual dying out of the species, but much, on the contrary, which points to an abrupt end, and evidence only of relation, not of actual time; while the recent valley deposit, although often indicating considerable age, shows rates of growth which, though variable, appear on the whole to have been comparatively rapid. The evidence, in fact, as it at present stands, does not seem to me to necessitate the carrying of man back in past time so much as bringing forward of extinct animals towards our own time; my own previous opinion, founded on an independent study of the superficial drift or pleistocene deposits, having likewise been certainly in favour of the latter view. There are numerous phenomena which I can only consider as evidence of a sudden change and of a rapid and transitory action and modification of the surface at a comparatively recent geological period,—a period which, if the foregoing facts are truly interpreted, would seem nevertheless to have been marked before its end by the presence of man on a land clothed with a vegetation apparently very similar to that now flourishing in like latitudes, and whose waters were inhabited by testacea, also of forms now living; while on the surface of that land there lived mammalia, of which some species are yet the associates of man, although accompanied by others, many of them of gigantic size, and of forms now extinct' (p. 248).

no support from the appearance of the earliest skulls which have been discovered, which, although evidently belonging to men who were driven to seek their sustenance in parts remote from the first seat of life, and among ferocious wild beasts, are yet, at least some of them, of a superior type, superior to those of their successors.[1] When man became savage it could only have been the result of degeneracy, and that this was so is abundantly accounted for on the reasons assigned by the Duke of Argyll.[2]

4. Hence we should expect to find that the first men, or the early races, were long-lived and powerful, that they followed peaceful occupations, were cunning workmen, made great advances in arts and sciences, and with this expectation history and investigation accord. We are told of the patriarchal longevity,[3] of giant size, of the immediate cultivation of the land, and the tending of sheep and cattle, the building of cities, of the invention of musical instruments, of artificers in metals, of astronomical skill; and besides the remains of art dug out of the Assyrian mounds, we have standing to this day in Egypt, not only the ruins of ancient temples, but pyramids and obelisks, and masterly sculpture, and other

---

[1] See Dawson's *Fossil Men*, pp. 181, 189.

[2] *Primeval Man*, Part iv. The fact stated by Guizot (*Earth and Man*, § 104), that the races of the southern continents are inferior to those of the northern, tends to corroborate his Grace's views, while the civilisation implied in the works of the mound builders of America, mentioned in Wilson's *Prehistoric Man*, vol. i. chap. x., shows that a civilised race may be succeeded by a savage one.

[3] Mr. Dawson, referring to the skeletons (whether fossil or otherwise) discovered of primitive men, speaks of the indication 'they seem to present of an extreme longevity,' of 'a slow maturity and great length of life,' also of great strength, and of brain power. See *Fossil Men*, pp. 181, 189, 198, 202.

evidences of the power and skill of the early workers outrivalling the efforts of modern times. None of these can claim to be as much as 6000 years old, though verging on them.[1]

[1] Since the above was written a book has been published by N. Joly, Professor at the Science Faculty of Toulouse, called *Man before Metals*, in which the great antiquity of the human race is advocated. Singularly enough, M. Joly appears to be unacquainted with the writings of the Americans, at least makes no allusion to their works. In a section titled 'The Great Antiquity of Man proved by Egyptian Monuments,' this author, after referring to a statement by Alfred Maury, that on the banks of the Nile art and civilisation date from a time anterior to all history, says, 'The development of this idea, which strongly supports our own theory, forms the subject of a learned paper by this well-known academician, with which the readers of the *Revue des deux Mondes* are doubtless acquainted. In order to dissipate any legitimate doubt as to the great antiquity of the Egyptian people, and of their civilisation and their arts, it is not necessary to cross the sea, to go to Karnac, and to penetrate into its temple, four times as large as Notre Dame in Paris, although it was reserved exclusively for the devotions of the king. It was enough to visit the little temple of Philæ in the Champ de Mars, where the rich treasures of the Egyptian Exhibition (1867) were displayed to the eyes of all nations. Their artistic beauty and richness, and above all, the art displayed in the adorning of the sepulchres, were very remarkable. This is due to the fact that the inhabitants of the Nile, ever preoccupied with the idea of a future life, looked upon the tombs "as their true abodes throughout eternity"' (Alfred Maury). Here were exposed coffins in the form of mummies, entirely covered with symbolic figures whose colours have resisted the ravages of time, and two statues, the one of diorite, the other of green basalt, representing King Chafra or Chephren (the fourth king of the fourth dynasty, and the builder of the second of the great pyramids), statues so well preserved, one of them especially, that they appear to be fresh from the hands of the able sculptor by whom they were carved more than 5000 years ago.

'Art does not attain at once to that grace of line and truth of expression of which the face of Pharaoh, son of Ra, the Sun god (Chafra), offers us an example. Side by side with these statues of King Chaefra or Schaffra, we may place the wooden statue of one named Ra-em-ke, remarkable for its wonderful state of preservation, and also for its beauty as a work of art, unsurpassed by any of the Greeks, as we are told by a competent judge (M. Fr. Lenormant). This Ra-em-ke was the governor of a province during the fifth dynasty, that is to say, about a century later than

5. If man had lived for a hundred thousand years upon the Earth, the Earth would, long before historic King Schaffra. Lastly, the door of the Great Pyramid of Sakkara, now one of the most precious treasures of the Berlin Museum, formed part of a monument which, if it were really built, as it is generally believed, under the first dynasty, has withstood for nearly sixty-eight centuries the destroying hand of man and of time.

'Such figures terrify the imagination. Forty-nine centuries before the birth of Christ is a great age for a work of human hands, and above all, for a true work of art. *Neither India, Asia, nor Assyria have any relics of a time which approaches so nearly to the origin of humanity.* But that which is really overwhelming to the mind is to find at that date not savage tribes, but a powerfully constituted society, of which the formation must have required the lapse of centuries; a civilised people advanced in science and art, and in the knowledge of mechanics, capable of raising monuments of immense size and of indestructible solidity.'

Then, after referring to certain Egyptian jewels 'of unequalled finish and beauty,' of the time of Joseph, and of other articles prior to that of Abraham, and averring that 'there can be no doubt that Egyptian art, so perfect under the reign of Chephren and his successors, began by equally rude attempts,' and that 'No one can tell with certainty the number of centuries they must have passed through before attaining to so complex a civilisation,' he concludes by saying, 'The whole history of Egypt confirms our belief in the immense antiquity of the human race.'—*Man before Metals*, by N. Joly, 1883, pp. 30–34.

There are several circumstances in this statement which invite observation. M. Joly tells us that no other nation than the Egyptian have relics of a period so near the origin of humanity, and this period, although it terrifies his imagination, he does not put farther back than 6800 years ago. It required, however, a few centuries to attain so great a civilisation, and from this he jumps to the conclusion that the antiquity of the human race is immense. Whatever observation may be made on the logic of this argument, it would appear that M. Joly conceives that the Egyptians could only have attained to such perfection in art after a lapse, perhaps he means a great lapse, of time. But there is no proper foundation for such a conclusion. If man was, as we believe, created perfect, it would need no great lapse of time, especially during the times of patriarchal longevity, to attain through experience to great skill in art. Perhaps the skill then attained very soon far exceeded all that is now possessed. Indeed, the wonder is, that after the long course of degradation to which in successive generations man has been subjected, man should now possess so much capacity as he now enjoys. It

times, have been filled with inhabitants. An old writer[1] presents a table of the increase which would naturally result from generation to generation, and at the end of 933 years the result would be 805,306,368 as the number of those then alive. Darwin places it higher, for he says[2] that in less than a thousand years there would literally not be standing room for man's progeny. Many causes, however, concur to keep down population. It would never grow and be sustained in the extraordinary numbers which may be arrived at by arithmetical calculation. Still it is obvious that only a few thousand years would have sufficed to people the world adequately. We should expect, therefore, had the Earth been peopled for 100,000 years, or even say 20,000 years, whether upon the progressive theory or not, to find not merely well authenticated histories of the race during these early ages, but everywhere evidences both of the existence of man, and that in masses, with extensive traces of his intelligent labour, it might be buried, and to be dug up, but still there. Yet not only does it appear that the population even of the most favoured lands was scanty, but out of certain central districts all that can be discovered only evidences a few early inhabitants roving, it may have been at first either adventurously or in search of food, from their former homes, without the

exhibits a wonderful recuperative power in nature. As regards the antiquity of the Egyptian nation, there is one passage in Exodus which tells strongly against the supposition. It is contained in the 24th verse of the 9th chapter, in which the plague of hail is incidentally mentioned as so grievous 'as there was none like it in all the land of Egypt *since it became a nation.*' This is one of those little bits of indirect evidence which are sometimes as convincing as more direct testimony.

[1] *Numbers of Mankind in Ancient and Modern Times* (Edin. 1753), p. 7.
[2] *Origin of Species*, 6th ed. p. 51.

means and appliances of their forefathers, and perhaps rapidly falling into a savage degeneracy. But nowhere are traces of intelligent operations to be found, save in those parts of the world of which Assyria and Egypt may be said to be centres.

6. And it was somewhere in these quarters that man originated. As stated by Southall,[1] 'the best opinion among ethnologists is that the migrations of the human race (supposing one original centre) commenced from Central or Western Asia.' Here then we ought, if anywhere, to find traces of this vast antiquity. But it is not so. Not a vestige of the existence of man or his implements beyond a period of 5000 or 6000 years can there be found.[2]

7. Lastly, we come to the only historical, or only reliable historical record we possess, a record confirmed by external evidence at every turn—the Bible. And here it is incumbent to observe that while it is true that the Bible was not given to teach us scientific truths, yet we cannot discard it as a witness to historical facts, even although it may be found to be itself the real defendant in the cause. It may present itself as a foreign witness, wearing a foreign garb, and speaking a foreign tongue, and requiring, it may be, a skilled interpreter. The interpreter may, indeed, mistake the meaning of the witness, and we require therefore to receive what he says with caution, the more particularly if the witness be speaking, and that very briefly and meagrely, to what he does not specially appear to prove. But here we have in Genesis, the oldest book in the world, written by one who had been taught everything that

[1] *Epoch of Mammoth*, p. 3.   [2] *Ibid.* p. 5.

was known to the most learned of early ancient nations, written, too, within a period not very far remote from those patriarchal times when men lived long,[1] and could transmit the knowledge of pristine times to a posterity through whose generations Moses could count his descent by recorded names. And if the interpreter be not mistaken as to the witness' meaning, we have here clear evidence of an antiquity for man which does not extend back from present times to more than 6000 or 7000 years.

But for the purposes of the present question it really is not of vital import whether man has lived 6000 or 60,000 or 600,000 years upon the Earth. The longest period which antiquity hunters could possibly name brings us to a time when man did not exist upon Earth; and in the history of our planet, which is by many reckoned by hundreds of millions of years, to say nothing of the history of creation, that period at the longest is comparatively short. Still it is right to think of it, as it truly ought to be thought, as not extending to 10,000 years, and really to a much shorter era, for it adds point to observation. Now if any Intelligence had visited this earth, say ten thousand years ago, he would not have found a single rational being existing on it. At that time, therefore, *the analogy of this Earth would have led to the conclusion that no rational being existed on any of the other planets*—that the worlds, indeed, were not made for rational life. But this ten thousand years is a mere point in eternity; and because what may be

---

[1] 'Before the flood men are said to have lived five, seven, nine hundred years; and as a physiologist I can assert positively that there is no fact reached by science to contradict or render this impossible.'—*Longevity*, by John Gardiner, M.D., p. 150.

represented as a speck or point in eternity has on our planet been blessed with rational life, is it to be inferred that other planetary bodies have been endowed with such life? Nor does it affect the consideration that for a long period, it might be millions, perhaps many millions, of years previously, organic life which was not rational, whether as mammal, or bird, or reptile, or fish, had dwelt on the earth or inhabited its waters. The fact only throws us (with the difference of excluding rational life) back a very little. Eternity remains unaffected; the speck is hardly bigger, if at all. Travel back to the time when the first vegetable form appeared, and the same observation applies.

Now this undoubted fact goes, I think, a long way, after all that has been already advanced; so long a way that I think I am almost fairly entitled to say that no presumption whatever arises from analogy to reason for life elsewhere because of present life upon the Earth.

Such a view is rendered the more reasonable when it is considered what life, and especially rational life is. Let us only look realizingly at it, and we shall be filled with amazement.

How wonderful is vegetation! The insignificant seed cast into the ground germinates, bursts, sends down its root, and sends up its shoot, and by an operation which we cannot comprehend, or even watch, slowly adds to both, till the root enlarges, and, piercing the ground, draws from it nourishment, and the shoot swells and rises little by little in gradual invisible progress, and grows and expands in stalks and trunk and leaves, beautifully veined, to draw nutrition from the air; and so, by imperceptible extension, addition after addition issues from the daily increasing stem, till the little shoot becomes a great tree, and the birds of the air lodge in its branches, while round and round, embosomed among the foliage, the flower unfolds, and, in graceful form and enlivening colour, adorns the waving boughs, and having served its day and brightened its generation, withering, gives place to fruit, pleasant to the sight and good for food both for man and beast, and to seed to reproduce its kind.

But proceeding to the higher kingdom, and passing by the inferior animal world, with all its interest and mysterious arrangements and adaptations, think of Man, who offers to consideration the most wonderful object of thought manifested in creation.

For he is wonderful in very many different ways. Wonderful in capacity to move about at will,—wonderful in the faculty of perceiving, through material, fleshly organs, surrounding nature and its accompaniments,—wonderful in that harmonious and complete combination of beautiful contrivances for the support and delight of existence exhibited in his bodily frame,—wonderful in ability, in accordance with his own desire, to exercise power over the external world, bending the forces of nature to aid him in his operations,—wonderful in the possession of the emotional sensations, in the impulse to love and to fear and to hate, to rear, to invent, and to destroy,—wonderful in the endowment of thought, the capacity to comprehend and mentally combine, to remember, to imagine, to reason, to deduce, to convince,—wonderful in the gift of speech, whereby to communicate the thoughts passing within his own inner consciousness to the inner consciousness of other similar living beings:

and, not suggesting other traits calculated in like manner upon reflection to excite astonishment, and leaving out of view what does not appeal so vividly to the outer observation, that most mysterious and most amazing characteristic of all, the carrying about within him of a spirit upon which its great Maker has breathed immortality, and adapted it to dwell with Him for ever,—how wonderful is the gift—that lofty faculty which raises him so far above every other terrestrial creature—to know, and the power to revere, the Being from whom he derived his life and soul! When we think of all, how forcibly do the Psalmist's words strike conviction to our hearts, we are 'fearfully and wonderfully made'!

Now mere matter is not so endowed. There is no love or hate, there is no moral consciousness, or consciousness of any kind, there is no intelligence, no knowledge, no power of recognising and worshipping God in a stone or any inorganic thing, nor in any inanimate organic thing. It is only when life is superadded that these qualities are found, and when life is withdrawn they disappear. The corpse, the skeleton of the dead, is as unconscious as the stone. Nor can man of himself, with all his bestowed powers, and still less can the inferior animals, impart to matter one of the properties or virtues named. Were it not that life is so common, so familiar to our experience, we should (if we can suppose the possibility of our looking upon the Earth with merely spiritual eyes) be astonished beyond measure at its feeblest manifestation. But because we are so habituated to it in its many forms, are so conscious of it within ourselves, we ought not to reckon it the less an exceeding great, a mysterious marvel. It is, in short, when rightly considered, in one sense of the term, a miracle. It is such a miracle on this Earth. What reason have we to suppose the miracle has been repeated elsewhere ? Literally none.

Ex nihilo nihil fit. The stone, the senseless atom, cannot communicate what it does not possess. Matter cannot of itself produce what is grander than itself. It has no life, and cannot give life. Life must be imparted to it, directly or indirectly, by Him who is Life, the great and only Being who is the fountain of life. Therefore the mere existence of other bodies precisely similar to the Earth (if the similarity can be established) affords no sufficient reason to suppose that the miracle of life has been repeated in them. Unless the Creator has so willed, this miracle has not been produced.

The difficulty of the introduction of life upon the Earth without miracle has been in various ways attempted to be met. Sir William Thomson some years ago suggested that life might be brought to this world on the tail of a comet, which was treated as a piece of Scotch humour (which probably it was), or in meteors ; and it has been thought that some of the aerolites showed traces of organic matter abiding in them. But examination (see *postea*) has disproved this supposition, and besides, it is most evident that any life which by possibility could have existed in these aerolites, whether animal or vegetable, would most effectually be burnt out upon entering the earth's atmosphere. The idea at one time entertained, that life can be initiated by ' spontaneous generation,' experiment has disproved.[1] M. Pasteur has

---

[1] The *Edinburgh Review* of April 1867, in an article on spontaneous generation, places at its head the following works : 1. ' Sur les Corpuscules Organisés qui existent dans l'Atmosphère, Examen de la doctrine des générations spontanées. Par M. L. Pasteur ' (*Annales de Chimie et de Physique*, tome LXIV. 1862). 2. *Traité de la génération spon-*

shown it to be untenable; and the experiments of others subsequently have apparently set the matter now at rest.

As little can the theory of evolution stand the test, and explain the production of life upon the earth, and especially the life of the rational being Man, without miraculous intervention. Evolution is a popular conception of the present day, just as in times past we have had notions of philosophers which, acquiring popularity, have lasted a transient existence; and notwithstanding the overbearing confidence of some naturalists, in whom the wish may be father to the thought, this new notion possibly will not endure any longer. For it is a mere unsupported *hypothesis*, and in any view it is incapable of explaining the presence or *origination of life* in this world. Let us, however, see upon what basis it rests, and what bearing the subject and the

tanée. Pouchet, Paris 1859. 3. 'Rapport fait à l'Académie des Sciences sur les expériences relatives à 4a génération spontanée' (*Annales de Chimie et de Physique*, tome IV. 1865). 4. 'Further Experiments on the Production of Organisms in closed Vessels. By George Child, M.D." (*Proc. Royal Society*, 1865). This article reviews the position of the investigation as at its date, and, stating the difficulty of proving a negative, leans to the conclusions of M. Pasteur.

Three years later Dr. Bastian experimented, and thought he had made out a case of spontaneous generation; but was promptly met by other experimenters, and it was shown that his results had been obtained by failure to take due precautions. The whole subject has been discussed at considerable length by Walter Noel Hartley, Demonstrator of Chemistry, etc., King's College, London, in his work, *Air and its Relations to Life* (1875). Mr. Hartley, who himself experimented with the utmost care, in a very interesting series of chapters (p. 132 et seq.) gives the history and state of the subject, confutes Dr. Bastian, and shows that the life supposed to be spontaneous is derived from germs floating in the air, and with which the air everywhere, except in higher regions, appears to be charged.

thoughts to which it gives rise have on the conclusions to be reached in the present question.

Evolution is not exactly a new doctrine. Lamark in 1801 broached the dogma of progressive development by acts of volition on the part of the animals themselves (a phase of the theory now given up, for the phases seem to be always shifting), and the descent of all species from other species. Similar views were stated by others, although the subject was not brought into public notice until 1844, when the author of the *Vestiges of Creation* first published that remarkable work, containing the idea that an original impulse was given to the forms of life aided by the influence of external circumstances, which advanced them through grades of organization. But Mr. Darwin, fifteen years later, in treating of 'the origin of species,' practically made the subject his own, so that Darwinism and Evolution, though not coincident, are commonly held to be convertible terms. His great theory to account for the diversity of species was that it was due to what he termed 'natural selection,' in stating which, or 'the survival of the fittest,' as his own, he appears to have met with claims of earlier origination on the part of Mr. Owen and Mr. Wallace. His conclusion was that all animals and plants are merely varieties of other and former forms, and by the operation of chance influences existing in nature through a long course of ages—one individual acting on another—have all been developed from one common ancestor. Mr. Darwin has, indeed, with great candour confessed to many of the weighty difficulties which beset his theory; at the same time endeavouring, with laboured ingenuity, to combat them by bringing 'natural selection' to his aid, so that the impression is given throughout of a *petitio principii* or of a reasoning in a circle.

His opinions were assailed by many, and among others, very shortly after the issue of the book, by Dr. Bree, in a work now perhaps not so much known, called *Species not Transmutable nor the Result of Secondary Causes*, where many telling facts and circumstances are adduced to show that Darwin's positions are untenable; Dr. Bree's declared object being [1] ' to show that Mr. Darwin's case is not proved, and that consequently it cannot supersede that theory of special creation of which we have the most convincing proof.' An attack in some respects more formidable, inasmuch as it was more systematic, and that it emanated from a friend and believer in Evolution, was afterwards made by Mr. St. George Mivart in his *Genesis of Species*. Mr. Mivart took up successively the different positions maintained by Mr. Darwin, and showed conclusively their error. But, after demolishing Mr. Darwin, Mr. Mivart failed in making out his own view, which is, that species have, in sudden and considerable changes, been evolved by ordinary laws (for the most part unknown), aided by the *subordinate* action of natural selection. And in a loose way adding,[2] ' there is nothing in physical science which forbids them (his readers) to regard these natural laws as acting with the divine concurrence, and in obedience to a creative fiat originally imposed on the primeval cosmos in the beginning by its Creator, its Upholder, and its Lord,' or, as Mr. Mivart elsewhere expresses it, ' throughout the whole process of physical evolution *supernatural action is not to be looked for.*' So that the distinct issue raised

---

[1] Bree's *Species not Transmutable*, p. 14.
[2] Mivart's *Genesis of Species*, p. 333.

by Evolution, so far as the present question is concerned, seems to be: *Are the separate forms of life the result of miracle?* that is, are they produced under the immediate act or direction of God,[1] evidenced by their being accomplished either by the use of means in an extraordinary manner, or without apparent means by the exercise of Divine Will, the manner of the accomplishment being here immaterial.

[1] I do not pretend to have read the multifarious writings which have been published on Evolution, and in dealing with it in this incidental way I have, availing myself in part of some of the observations made by other writers, taken my own method,—others have taken different methods. But one thing is noticeable, that Evolutionists are very anxious that it should not be supposed that their doctrine militates against religion. Yet no one can intelligently read the works of Evolutionists without observing how carefully many, if they do not openly seek to eliminate the action of the Creator, evade acknowledging it, and, when anything is reluctantly said on the subject, the impression is conveyed that it is so said as it were to gain the suffrages of religious people, or not altogether to shock them. Principal Dawson thus expresses himself (*Story of Earth and Man*, p. 317): 'This Evolutionist doctrine is itself one of the strangest phenomena of humanity. It existed, and most naturally, in the oldest philosophy and poetry, in connection with the crudest and most uncritical attempts of the human mind to grasp the system of nature; but that in our day a system destitute of any shadow of proof, and supported merely by vague analogies and figures of speech, and by the arbitrary and artificial coherence of its own parts, should be accepted as a philosophy, and should find able adherents to string upon its thread of hypotheses our vast and weighty stores of knowledge, is surpassingly strange. It seems to indicate that the accumulated facts of our age have gone altogether beyond its capacity for generalization; and, but for the vigour one sees everywhere, it might be taken as an indication that the human mind has fallen into a state of senility, and in its dotage mistakes for science the imaginations which were the dreams of its youth. In many respects these speculations are important and worthy of the attention of thinking men. They seek to revolutionize the religious beliefs of the world, and if accepted would destroy most of the existing theology and philosophy. They indicate tendencies among scientific thinkers which, though probably temporary, must, before they disappear, descend to lower strata, and reproduce themselves in grosser forms and with most serious

First, then, with regard to the Darwinian theory, upon which these observations among others occur.

1. It is at once conceded that species are 'under the Divine concurrence' subject to variation, in some cases effects on the whole structure of society.' On the other hand, Mr. George St. Clair, who seems to have given himself over, body and soul, to the Evolutionists, and, if I have not mistaken his meaning, to contend for *The Vestiges* doctrine of original fiat out of which everything has sprung, has written a book, *Darwinism and Design; or, Creation by Evolution* (1873), bearing that his endeavour is to 'show that if Evolution be true all is not lost, but on the contrary something is gained; the Design argument remains unshaken, and the wisdom and beneficence of God receive new illustration.' But obviously it is a truism to say that if Evolution be true it has been the result of design. The antecedent question must first be answered. Is it true? Is it satisfactorily proved? There is no escape from this position. If it were true, it would clearly form part of the Creator's plan. But to attribute to Almighty design seems no part of the Darwinian doctrine. Dr. Charles Hodge, in a work called *Darwinism and its Relation to the Truths of Natural and Revealed Religion*, enters at some length into the theology of the question, and may be consulted as stating the views of many writers, and thus expresses himself (p. 173): 'The conclusion of the whole matter is, that the denial of design in nature is virtually the denial of God. Mr. Darwin's theory does deny all design in nature, therefore his theory is virtually atheistical; his theory, not he himself. He believes in a Creator.' [How far does this accord with his lately published letter?] 'But when that Creator, millions and millions of ages ago, did something—called matter and a living germ into existence —and then abandoned the Universe to itself, to be controlled by chance and necessity, without any purpose on His part as to the result, or any intervention or guidance, then He is virtually consigned, so far as we are concerned, to non-existence. It has already been said that the most extreme of Mr. Darwin's admirers adopt and laud his theory, for the special reason that it banishes God from the world; that it enables them to account for design without referring it to the purpose or agency of God. This is done expressly by Büchner, Haeckel, Vogt, and Strauss. The opponents of Darwinism direct their objections principally against this element of the doctrine. This, as was stated by the Rev. Dr. Peabody, was the main ground of the earnest opposition of Agassiz to the theory. America's great botanist, Dr. Asa Gray, avows himself an Evolutionist; but he is not a Darwinian. Of that point we have the clearest possible proof.'

to a pretty wide degree. And no doubt this is a susceptibility which has been implanted by the Creator not merely to manifest His own wonderful power, and to gratify His own grand love of multiplied diversity, but for various good ends, and especially the better adaptation of animals to changed and surrounding circumstances, and, above all, both the better to minister to the wants of man, to whom the whole inferior creation was subjected, and the more to stimulate his energies and quicken his intelligence.

2. But this susceptibility, which may be regarded as part of the Divine mandate, 'Be fruitful and multiply,' has its limits. According to Mr. Darwin's doctrine, the lion and the lamb, the crocodile and the ape, the elephant and the animalcule, the spider and the whale, the sparrow and the shark, the centipede and the man, the owl and the ass, the goose and the gull, have all sprung from one common ancestor, all arriving at their respective forms by minute, gradual, successive changes. This is a community of origin which obviously cannot claim for itself the smallest shadow of *probability*. But what are the facts?

(1.) The different races of animals keep themselves apart. They herd together; they pair together. We see no disposition on the part of two different kinds of animals to form unions with each other. There may be, and when it does happen it is only noticeable for its singularity, an affection springing up between two animals of different genera; but it is purely Platonic.

(2.) Even animals which may be regarded as kin to each other seldom pair. It happens among domestic animals, or animals which are directly useful to man, and have been brought under his mastery, and therefore conceivably created with this special power of adaptation; but it has only been observed while in a state of domesticity, and therefore under the control and guidance of man; and it would almost seem to be admitted by the Evolutionists that among wild animals, or animals in a state of nature, it does not happen, or if it does happen, that it can only be in the rarest cases.

(3.) There is, it seems, as stated by Darwin himself, corroborated by Mivart,[1] a tendency to revert in time, after passing through the change of variation, to the original typical form.

(4.) When pairing takes place between animals of different genera, the offspring is as a rule sterile.[2] The conditions under which absolute sterility does not follow may be regarded as exceptions proving the rule.

(5.) There appears to be an increasing 'difficulty found in producing by ever such careful selection any further extreme in some change already carried very far, such as the tail of the "fan-tailed pigeon" or the crop of the pouter.'[3]

(6.) There are some animals, such as the goose, the peacock, and the guinea fowl,[4] which are almost incapable of variation, and others[5] which have remained unvaried since the time of their creation.

In short, every circumstance goes to show what Mr. Mivart calls the 'stability' of species;[6] and even Mr. Darwin seems to admit that in the great divisions of the animal kingdom it is impossible that the one can become the other. 'Mammals and fish hardly come into competition with each other; the advancement of

---

[1] *Origin of Species*, 6th edition, p. 11; *Genesis of Species*, p. 137.
[2] Darwin, p. 241.   [3] Mivart, p. 131.   [4] *Ibid.* p. 134.
[5] Darwin, p. 308.   [6] Mivart, p. 127.

the whole class of mammals, or of certain members in their class, to the highest grade, would not lead to their taking the place of fishes.'[1] In fact, is it not that the various branches of the animal kingdom have obeyed the original Divine order, 'Let the Earth bring forth the living creature *after his kind*, cattle and creeping thing and beast of the Earth *after his kind*: and it was so.' And so beautifully separated into kinds by distinguishing characteristics are they, that while naturalists have arranged the whole according to peculiarities into scientific systems, in which the animals are classified into distinct genera, which again are subdivided into varied species, the partitions, in at least their greater divisions, are so marked as to be palpable to every eye, from that of our great progenitor who called each kind by its individual name as it passed before him, to that of the tiniest nursery child of the present day.

3. There are many facts in natural history adduced by naturalists which are irreconcilable with Mr. Darwin's theory; as, for example, the existence of the same fresh-water fish in fresh-water lakes in different regions of the Earth,[2] and the possession of perfect eyes and ears[3] in animals living in the earliest times[4]— organs which 'could never have been produced by insignificant beginnings.'[5] And of another kind, the fact mentioned by the Duke of Argyll in regard to the specific variation in the humming birds—

'That every variety which is to take its place as a new species must be born male and female; because it is one of the facts of specific variation in the humming birds, that although the male and female plumage is generally

---

[1] Mivart, p. 99.   [2] P. 165.   [3] P. 152.
[4] P. 84.   [5] P. 58.

entirely different, yet the female of each species is as distinct from the female of every other as the male is from the male of every other;' with which the further fact has to be coupled that there is no 'hybridism between any two species however nearly allied.'[1]

4. The apparently rapid succession of races or species in geologic times is opposed to the Darwinian theory.

Discussing the succession of life upon the globe, and in dealing with Evolution, Dr. Nicholson says:—

'On the other hand, there are facts which point clearly to the existence of some law other than that of Evolution, and probably of a deeper and more far-reaching character. Upon no theory of Evolution can we find a satisfactory explanation for the constant introduction throughout geological time of new forms of life, which do not appear to have been preceded by pre-existent allied types. The Graptolites and Trilobites have no known predecessors, and leave no known successors. The insects appear suddenly in the Devonian, and the Arachnides, and Myriapods in the Carboniferous, under well differentiated and highly specialized types. The Debranchiate Cephalopods appear with equal apparent suddenness on the older Mesozoic deposits, and no known type of the Palæozoic period can be pointed to as a possible ancestor. The Hippuritidæ of the Cretaceous burst into a varied life to all appearance almost immediately after their first introduction into existence. The wonderful Dicotyledonous flora of the upper Cretaceous period similarly surprises us without any prophetic annunciation from the older Jurassic.'[2]

5. Evolution from a mere sentient existence cannot explain the possession of passions, such as love or hate; or moral feelings, such as the sense of right and wrong;

---

[1] *Reign of Law*, p. 237.
[2] *The Ancient Life History of the Earth*, by H. Alleyne Nicholson, M.D., p. 373.

nor can it explain in man, what is absent in the brute, the conception and reverence of God, and, without putting it higher, his hope of immortality.

6. Mr. Darwin admits as a serious difficulty, started by Sir W. Thompson, ' probably one of the gravest yet advanced,'[1] that the lapse of time since our planet was consolidated is not sufficient for the amount of organic change to be produced, according to his theory, by minute succession and favourable variations—a difficulty in which Professor Huxley seems to participate. For, according to estimates formed on the principles of inquiry followed by Sir W. Thompson, life on the Earth must be limited within some such period of time as one hundred million years.[2] Whereas, according to computations founded on the Darwinian theory, it would require 2500 millions of years[3] for the complete development of the whole animal kingdom to its present state, even were such development by natural selection possible. These are calculations which, supposing the theory well founded, are necessarily speculative, and both of them might be largely either too great or too small. But the least reflection will show that upon the Darwinian supposition of the formation of animals of the most opposite kinds—say of the Megatherium and the Ammonite of the early world, or the elephant and the lark of the present time, or of the giant whale and the animalcule (one million to the bulk of a grain of sand)—all from the Eozoon, or whatever was the first created animal form, and that by minute successive changes accidentally happening and long in maturing, would, were it possible, require an inconceivable length of time.

[1] *Origin of Species*, p. 409.   [2] Mivart, p. 154.   [3] P. 160.

7. Further, it is found that existing animals have lived on Earth back to the glacial period without change. It seems that 98 species of mammalia inhabited Europe in the post-glacial period, of which 57 still exist, not one of which has been modified into a new form. Principal Dawson, in mentioning this fact, adds [1] that he had examined about 200 species of molluscs and other marine animals, some of which lived back to the Tertiary period, and he had arrived at the conclusion that they are absolutely unchanged.

8. But a more tangible and formidable, nay, insuperable objection is to be found in the admitted fact that there is an entire absence of all proof—where proof ought to have been abundant—of the theory. It would, indeed, if well founded have been *a matter of everyday observation.* Yet we neither see such changes happening before our eyes, nor do we read of them in the history of the past. While there is not a trace of them such as is requisite to be found in the geological records, where, had the race of animals now or formerly existing been produced by the development supposed, there must necessarily have been preserved to us specimens of animals in all the intermediate or transitional stages. Mr. Darwin admits[2] that the number of these 'intermediate varieties which have formerly existed must be truly enormous,' and that 'geology assuredly does not reveal any such finely graduated organic change; and this, perhaps, is the most serious and obvious objection which can be urged against the theory. The explanation,' he adds, 'lies, as I believe, in the extreme imperfection of the geological record.'[3] But this is

---

[1] *Story of the Earth and Man*, p. 358.
[2] *Origin of Species*, p. 264.　　　　[3] P. 265.

an explanation of a fact so strongly against him which cannot be taken. The simple answer to it is, to use Mr. Mivart's words, that 'the absence not in one, but *in all cases* of such connecting links, is a phenomenon which remains to be accounted for;'[1] but is it not conclusively adverse to the theory?

The more recent Evolutionists, it is true, have taken courage from finding, not indeed an 'enormous number of intermediate varieties,' but a few isolated individuals, affording what they consider to be examples of the 'links' of which they are in quest. Some of them, if they are to be assumed the results of Evolution, would, no doubt, be cases of that development of which, as already stated, species are or may be susceptible. Others are thought to typify a class exemplifying connection between different races or genera, and of these the Archæopteryx is paraded with triumph as furnishing *probatio probata* of the Darwinian theory. This singular animal, which united the feathers and certain limbs of the bird to the structure of the reptile, and has therefore been set up to constitute a link by Evolution between the reptile and the bird, may doubtless be regarded as a product of a period of transition between two geological epochs; the epoch which was favourable to reptilia, and that newer epoch in which the atmosphere became suitable for the bird; and, no doubt, it was created in that particular form as well for its own preservation as for the purpose of enabling it the better to pursue its peculiar prey, and otherwise fulfil its place in creation. We may rest assured that such an animal was needed, and that the special form of this to us seeming *lusus naturæ* was what was

[1] Mivart, p. 125.

best adapted for the purpose for which it was introduced.

But, dealing with the matter as a scientific question, a far stronger reason, and one applicable to 'links' of all kinds, is this, that even could it be established that they were *related in time*, it must also be established that they were the *product of Evolution*. The mere fact of correspondence or similarity is obviously insufficient. The one must be proved to be the parent of the other, and this ought to be demonstrated by evidence either direct or circumstantial, though sure and relevant. Now it may be admitted that in the circumstances neither is possible. The question, therefore, can only be determined, if at all, by reasoning from analogy,—from the known and visible facts of the present we may reason back to the unknown and hidden secrets of the past. But so far from finding the known and visible to afford evidence in favour of the theory, we have already seen that all the facts are opposed to the hypothesis. Hence it finds as little support from analogy as from evidence.

There exists, therefore, abundant reason for putting aside the Darwinian theory; and we may now turn to the other supposition propounded by Mivart, and which receives no countenance from the views of Darwin,[1] and see whether it be any better—that the changes have been sudden and considerable, and 'have been evolved

[1] 'It may be doubted whether sudden and considerable deviations of structure, such as we occasionally see in our domestic productions, more especially with plants, are ever permanently propagated in a state of nature. Almost every part of every organic being is so beautifully related to its complex conditions of life, that it seems as improbable that any part should have been suddenly produced perfect, as that a complex machine should have been invented by man in a perfect state.'—*Origin of Species*, p. 33.

by natural laws (for the most part unknown), aided by the *subordinate* action of natural selection.'[1]

Mr. Mivart, like all Evolutionists, proceeds on the footing that similarity of parts is evidence of affinity; and his position is, that the production of this similarity, or what has been termed the 'homologies' of the limbs, is the result of '*an internal force or tendency*,'[2] without excluding the contributing action or effects of 'external conditions,' but excluding all supernatural action of the Creator.

On looking, however, for the proofs, it will be found that the whole rests on an *assumption of the doctrine of affinity because of resemblance of parts* or members, or, in other words, of Evolution,—which is nothing more than a complete begging of the question,—and, basing his remarks on this footing, that it is '*inconceivable*' that indefinite variation with survival of the fittest can ever have built up these serial, bilateral, and vertical homologies '*without the action of some innate power or tendency so to build up possessed by the organism itself in each case.*' This is clearly a method of argument which is false; and true proofs of this theory there are none. Indeed, he himself[3] is forced to rest on assertion, and to admit that the 'balancing' which results in the given phenomena 'must be due to an internal cause, which at present science is utterly incompetent to explain.'

Now we know that the Creator usually works by means or instrumentalities, and therefore it is not at all incredible that He should implant in His creatures a power which, acted upon by Him, and intensified by

[1] Mivart, p. 333. [2] P. 185. [3] P. 207.

Him, and directed by His infinite knowledge for His own purposes, would result in a rapid change of form, or a change of form in descendants. But while this method *may* have been adopted by Him in some cases,—although there is nothing to show that it *has been* adopted,—we cannot suppose it to have happened, in any case, *unless He had directly so willed and ordained*, and had given the impetus to produce in this way and in the specific forms; but this assuredly would have been a miracle, in the sense of being that supernatural interposition of God in the affairs of this world which every devout reader of his Bible is accustomed to contemplate, and to believe without difficulty.

But if a miracle be needed, it can be exerted by the Creator in any way He sees proper. His purpose may be effected by the exercise simply of His will. It may be brought about, in a way we know not, by animating the dust; for 'He can animate the dust as easily as make the dust itself;' or, as He did at the first, He can create out of nothing. People forget, when they speak of a miracle, that everything around us is not only marvellous, but incomprehensible. The rain that falls to the ground mounts, by a wonderful process, into the vine, and becomes the juice of the grape, which, subjected by man—employing the forces imparted to matter by the Creator—to certain processes, becomes wine. But, attentively considered, this result is not in reality a whit less wonderful than our Saviour's exerting His Almighty Power and at once, and by the exercise of His will,—without the means which the forces of nature had derived originally from His Power,—turning the water into wine. These forces of nature came from Him. They were nowhere till He implanted them; and

He who could originate and implant them, and was thus above them and could control them as He pleased, could surely accomplish the same end, without their aid or instrumentality, by the exercise of the same Power which brought these means and forces out of nothing and implanted them in matter.[1]

The action of man, the creature,—who derives all his finite powers from the Creator, and who can only use the means which the Creator has provided, but who can even influence by a look,—is admitted; and yet the action of God, whose power is infinite, is excluded from the world He has Himself framed. Man can alter the face of nature,—man can do many things in exercise of his will; but surely what can be done by man can be done, and much more abundantly, by the Almighty, and that by ways and means which to man are unknown.

I have put it as if the creation of the animal world was accomplished by the mere exercise of will; for it is well to keep in view that the Creator *can* so act. 'That God is not limited to ordinary law in the production of results, that He can and that He does produce events without the ordinary instruments of nature, is,' says Henry Ward Beecher, 'the very spirit of the whole Bible.'[2] But it by no means follows that the creation was, at least in all its parts, accomplished without what we call means. It may have been that there are special laws of creation of which we know nothing, or special modes of operating upon known forces which produce

---

[1] I have not ventured here, where it would be out of place, into the general question of miracles, but only said sufficient logically to reach my conclusion applicable to the present question.

[2] Beecher's *Life of Christ*, p. 36.

creation. But then this special law, or these special forces, in the very nature of the case, will never be brought into operation until the Divine Being will and direct, and the evidence of such direction lies in the beautiful marks of design manifested everywhere in creation.[1] The adaptations we see everywhere around us, the exquisite contrivances (so well exemplified by the noble duke[2] in the wing of a bird), indicate, in the plainest manner, design and the hand of a Designer; and they are not fashioned in an iron and rigid form, —are not turned out of a common mould, with that identity of construction which a machine, subject to no control, put in motion would produce,—but are manifestly obedient to a guidance which shapes them to the immediate requirement; and when they obey external influences, these are influences which have only derived their power from the Creator Himself.

The Evolutionists have found it necessary to their doctrine to carry it the length of including man. Now it might have been that the inferior creation was, at least to a large extent, derived from other of the subordinate forms, and yet it would not follow that man, so specially made the subject of the Divine decree, 'Let us make man in our image,' was evolved from a brute.

That man was descended from the brute creation is, however, so much a dogma of Darwin, that, in his *Descent of Man*, we are favoured with the exact description of the human ancestry. 'The early progenitors

---

[1] The evidences of design would fill volumes. Take, for example, the elephant's trunk. Cuvier found that in it the number of distinct muscles, each having its separate action, is not far short of forty thousand. Or take the beautiful, the marvellous and minutely exact adjustments everywhere manifested in the solar system.

[2] *Reign of Law*, chap. iii.

of man were no doubt once covered with hair, both sexes having beards; their ears were pointed and capable of movement, and their bodies were provided with a tail having the proper muscles.'[1] 'The foot was prehensile, and our progenitors were no doubt arboreal in their habits.' And as, according to the Evolution theory, we must have sprung from the lowest form of life, he tells us: 'At a still earlier period the progenitors of man must have been aquatic in their habits.'[2] And, as if 'the art of sinking' could no farther go: 'These early predecessors of man must have been as lowly organized as the lancelet or amphioxus, or even still more lowly organized.'

It is to the monkey Evolutionists proximately ascribe their descent.[3] But in so contending, there is one staggering fact which is ignored by most of them, perhaps is not known to many. Professor Owen, who considers the gorilla to be the most anthropoid ape known (and it will be borne in mind it is the habitant of equatorial Africa, far removed from the received birthplace of the human race), has collected the differences which distinguish that species[4] of monkey from man, amounting in all to 44 separate peculiarities, leading him to the conclusion that man forms one species.[5] But

---

[1] *Descent of Man*, p. 206.   [2] P. 207.

[3] So recently as at a meeting of the British Association held at Southampton in September 1882, this view was contended for by Mr. Stewart Duncan in a paper read to the Association. However, it is somewhat satisfactory to know from the newspaper report, that 'some members tried to throw over the ape theory;' although, it is added, 'Mr. Duncan held that if it could be shown man walked on all-fours, and was otherwise very low in the scale, then practically man was an ape.'

[4] See Dr. Bree's *Species not Transmutable*, p. 111.

[5] See also Dawson's *Story of Earth and Man*, p. 363 et seq. Principal Dawson shows not only distinctive differences between them and the

the mass of differences would, according to the Darwinian theory,—contrary to the simplicity characterizing the Divine modes,—have involved innumerable transitions and enormous time for its development from one to the other, till they culminated in man. Yet not a trace of these transitional forms can be discovered. The author of the *Vestiges* gave it up. 'It might now be desirable,' he said,[1] ' to trace the connections of that order in the upward progress towards the highest of animated beings; *but here such obscurity prevails that I must be content to leave the task to other inquirers.*' Darwin himself, in his work expressly treating of the descent of man,[2] is forced to speak of ' *the great break in the organic chain between man and his nearest allies, which cannot be bridged over by any extinct or living species;*' and all he can urge in favour,[3] is that many forms have become extinct,—a position altogether inadmissible in the face of the present state of historical and geological knowledge.[4] It might, indeed, be sufficient to remark, Why do we not

lower animals, but that man was truly a new and isolated creation,—a new type of life having been produced in him. I have not thought it necessary to advert to the discussion which has arisen as to the place man should hold in the zoological order. This seems to turn upon the structure of the brain and other anatomical considerations which, in the view I take, do not affect the question. See Lyell's *Antiquity of Man*, chap. xxiv.

[1] *Vestiges*, 4th ed. p. 272.   [2] Darwin's *Descent*, p. 200.   [3] P. 187.

[4] Mr. Huxley, who was at least a conditional Darwinian, writes to much the same effect : ' Every bone of a gorilla bears marks by which it might be distinguished from the corresponding bone of a man ; and in the present creation, at any rate, no intermediate limb bridges over the gap between *Homo* and *Troglodites*.'—*Man's Place in Nature*, p. 104. M. Paul du Chaillu, whose remarkable adventures among the gorillas at the time he published his famous book were much canvassed, but who probably has become better acquainted with that animal than any one before or since, discussing the physiological resemblances and differences between the gorilla and man at full length, observes : ' Though there

now, at the present day, observe apes becoming, or giving birth to, a human being, or any intermediate form?

But what has always appeared to me as lying at the root of the Evolution fallacy, is in imagining that resemblance of body, or limb, or embryo, denotes affinity. This idea pervades the whole system, and is present with every writer. Look, for example, at Huxley's book, *Man's Place in Nature*, with its grotesque skeleton procession, a book written to establish the resemblance of man to the monkey, and our descent from that animal because of such resemblance. Eliminate this element, and Evolution is nowhere.

Now when we see several buildings all marked by the same general pattern, or having one pervading cast, in ascribing them to a common designer we infer that it is his ideal, rightly or wrongly conceived, of the best or most perfect form for structure of the particular style or class to which it belongs; in building which he does

---

is a great dissimilarity between the bony frame of man and that of the gorilla, there is also an awful likeness, which in the gorilla resembles an exaggerated caricature of a human being. . . . While I resided among the tribes found on the mountains near the banks of the Ovenga river, where the gorilla is rather more common than anywhere else, I searched in vain if an intermediate race, or rather several intermediate races or links, between the natives and the gorilla could be found; and I must say here I made those inquiries conscientiously, with the sole view of bringing before science the facts which I might collect. But I have searched in vain: I found not a single being, young or old, who could show an intermediate link between man and the gorilla, which would certainly be found if man had come from the ape. I suppose from these facts we must come to the conclusion, *that man belongs to a distinct family from that of the ape,*—the first belonging to the order Bimana, and the latter to the other quadrumanous series.'—*Explorations and Adventures in Equatorial Africa*, by Paul B. du Chaillu, p. 378. John Murray, 1865.

not necessarily employ the same workmen, or extract the material from the same quarry. And so, though infinitely more, when we find in nature a large number of forms of life of a similar kind, having structural resemblance in common, the true inference is, that they all have proceeded from the hand of the one Great Designer, and that *He, in the perfection of His wisdom, has chosen that general form which is the most completely adapted for the purpose in creation which each animal of the kind was designed to subserve, and which was best calculated to sustain or otherwise minister to its existence; so that it could not be said that any other would be as perfect; and therefore to have employed any other would not have been in accordance with Divine wisdom.* When one animal proceeds from another, it is because of the Divine Power which has been communicated, for without the Divine Power the forces used would have been as inefficient as dead matter. But the resemblance of any two species does not necessarily establish that the one proceeds from the other; and really Evolutionists must think people extremely credulous, or extremely illogical, when they adduce their so-called 'links' to verify their theory. All that resemblance certainly proves, is not that the one proceeded from the other, but that the similar creatures have been originated by the one Great Being, and that that Being designs and fashions in that particular form which is the most perfectly adapted, in the time and place, for their subservient purpose.

Applying this view to man, we find him formed and endowed according to the most perfect model; and if in any part or member its form resembles that of the inferior animals, it is simply because the form is that which the Great Architect and Designer *considered to be, and which unquestionably is found to be, best adapted for both man and beast* to attain the end which the particular part was intended to fulfil. And so clearly was the introduction of man the effect of design, that he and woman (differing in many ways from each other, but the one being the true complement of the other), the male and female of the human family, were approximately coeval in their creation; without which cotemporary formation the race could not have been perpetuated unless by continually renewed miracle. It was also, or must have been, the same with the male and female of every other creature.

Mr. Wallace,[1] notwithstanding he is a pronounced Evolutionist, and a strong advocate of natural selection,—albeit his own statement of the striking adaptation of animals in form and colour, to enable them to elude their enemies, is a strong argument against his theory of survival of the fittest,—has found the facts for the separation of man to be so conclusive, that, regarding specially the size of his brain, the absence of hairy covering on the back, the structure of the hands and feet, his mental faculties, and the possession of the moral sense, he is driven to the conclusion (apparently) that man cannot be the product of natural selection; and (although he cannot shake himself free of the theory of development as the instrumentality) '*that a superior intelligence has guided the development of man in a definite direction and for a special purpose*, just as man guides the development of many animal and vegetable forms.' An admission which he unhappily mars by adding, 'I must confess that this theory has the

[1] Wallace's *Contributions to Theory of Natural Selection*, p. 52 et seq.

disadvantage of requiring the intervention of some distinct intelligence to aid in the production.' [1]

One would have thought that, so far from a disadvantage, it afforded the simplest method to solve a difficulty which can never otherwise be solved. But it implies supernatural intervention in the case of man; *and the admission of miracle in the case of man*, although he is the highest in rank of all, *necessarily gives room for the intervention of miracle in other or in all cases.*

Leaving, therefore, naturalists to do battle over details which they alone are competent to discuss,[2] and to account for a desire which some of them have to trace their origin to the ape and to still lower orders, rather than rise to the nobler thought, coming to us from a higher source and a superior authority, that the genealogy of man stopped short at Adam, 'who was the son of God;' and dealing with the matter on the broad principles which can be applied to a doctrine

---

[1] P. 359.

[2] For this reason I have taken no note of one of the latest works on the subject, Dr. Andrew Wilson's *Chapters on Evolution*, which is a book for the naturalist to discuss. I am afraid were I to attempt to enter upon it I should make as awkward a mistake as he has himself done in venturing into the demesnes of law to explain a law term, which has not without advantage been imported into logic, or rather into argument— the 'taking an objection to the relevancy.' This he expounds to mean, 'showing that certain parts of the statement of facts made by the opposing side involve items which may be absolutely untrue or incorrect, and which therefore require to be expunged from the list of matters involving litigation.' Now, on the contrary, an objection to relevancy assumes the truth of the averment, but denies that on this assumption it affects or warrants the conclusion. Just as if we were admitting all Dr. Wilson's statements of fact to be true, but saying they do not infer demonstration of the proposition he maintains. It would be well indeed if in all cases the relevancy or adequacy of statements to infer conclusions were duly considered.

which is only at best a hypothesis,—a hypothesis which must yield to facts, in place of our trying to bend the facts to it,—there is in reality nothing to warrant our regarding the unsupported theory as affording any explanation of the production of life upon the Earth by a process of development. On the contrary, examination of the subject shows that all life, and, above all, the existence of rational man, is due to the exercise of miraculous agency.

Indeed, the cessation of new production is both an evidence of miracle in the past and a miracle in itself, were there any truth in Evolution.

After all, Evolution does not explain the origin of life. As one of its advocates says, 'the history of development but environs the puzzles connected with life and its nature. It leads us to the beginnings of life, it is true, but it leaves these beginnings unaccounted for, and as mysterious as before.'[1]

The author of the *Vestiges* perceived the length to which he must carry his theory, if he maintained it at all; for the tracing back to one original form of life left the production of that form unexplained, and so he was driven to the doctrine of 'spontaneous generation,' and in the *Sequel* to his work he presented the inconclusive experiments of Mr. Weekes. But better investigation than that of Mr. Weekes, and more careful experiment, have entirely disproved the supposition; and it may clinch the argument to say that if all life and seeds of life on Earth were to be now destroyed, say by the precipitation of a planet into the Sun, or by our planet being brought into close proximity to the Sun,

[1] Wilson's *Chapters on Evolution*, p. 168.

so that every vestige of life should be burnt out (just as if, could it possibly have previously existed, it must have been burnt out in the primeval fires), life would not return; a miracle would be needed to replenish and repeople the Earth.

We thus start with this most important consideration, that life on Earth does not necessarily or spontaneously spring up, but is produced in all its manifestations—no matter how, for the manner of origination is immaterial—by the will and miraculous intervention of the Creator.

But we do not stop short here. For we find that the Earth has been peculiarly fitted for life, so that a reason existed for its receiving life. Placed at a medium distance from the great source of light and heat in the planetary system, the light and heat are of that exact degree which favour existence. They neither scorch nor chill; vegetation adapted to support animal life receives the exact measure which conduces to its growth. It possesses an atmosphere composed of the needful elements, in the precise combination which is calculated to maintain life, both animal and vegetable. It is supplied with water of a quality exactly suited to nourish and maintain life. Its surface is so planned and laid out in mountain, valley, river, and sea as to promote in the best manner these operations—by cloud, and frost, and wind, and everything else without which Earth would be uninhabitable. It has been so adjusted in distance from the Sun and Moon, in the extent and form of its oceans, and in the bounds set by the shore, that the tides do not rise to inundate the land; and the proud wave, carrying salubrity in its swell, is shut up within doors, and told hitherto shall thou come, and no farther.

Furthermore, the Earth was not left as a mere locality for life. All nature was put in operation to prepare it specially for man. After the great nebular gaseous globe, of which it at first consisted, had, in the course of incalculable ages, cooled down, and contracted to its present dimensions, and the hot, enveloping mist had broken up and cleared away, and changed into an aqueous covering, and the volcanic forces within had upheaved the great mountains, millions of years of preparation were needed to bring it by successive operations and succeeding races of animal and vegetable life, exactly suited to their time and place, into fitness for man; and from the minutest organism—from the earliest known form of animated existence, from the Eozoon,[1] discovered far down in the geologic depths, which multiplied and extended till it filled whole seas, and produced enormous tracts of rocks of chalk—to the gigantic monsters, the unwieldy Megatheria, the huge Mastodons and Mammoths, which were fitted to grapple with the luxuriant tropical forests equally gigantic,—all were set to work to make Earth a suitable and pleasant habitation for man, and having made it so the monsters were caused to disappear; and, evidencing the grand preparation plan, there were in their stead along with man, or shortly before, introduced 'several mammalian species not known in the Pliocene period, and some of which, as the sheep, the goat, the ox, and the dog, have ever since been his companions and humble allies.'[2]

And thus it is we find coal formed and stored in the far back ages for man to burn in the present age. But no other animal uses fire or digs for coal. And stone

---

[1] *The Dawn of Life*, where Dr. Dawson has devoted a book to the subject of Eozoon.

[2] Dawson's *Story of Earth and Man*, p. 289.

is found fashioned and in deposits slowly laid at the bottom of the ocean in remote epochs, and then elevated within man's reach by volcanic force to afford to man the materials with which to build and embellish; but no other animal builds or embellishes with stone. And the rare and valued jewel is found sometimes in the depths of Earth, and sometimes washed up by the sea; but no other animal knows or appreciates its charms, searches for it, cuts and polishes and sets in costly gold, or adorns its person with the sparkling brilliant, the finished gem. And metals are in due measure found streaking the Earth in veins and lodes, and have been given to man to aid his industry, augment his powers and resources, and to increase his knowledge; but no other animal sinks a shaft, raises the ore, eliminates the metal, forms it into shape, or uses a tool. And animal life has been given that he may use it in a way that he alone can use it—the furry skin, the soft wool, the silken cocoon, the tough hide— for clothing and other purposes;—the horse, the ass, the camel to ride, or to draw, or bear his burdens. While the vegetable world is also largely laid under contribution to meet the wants of man alone. Flax and cotton grow that man may, by the aid of invented machinery, manufacture dress; but no other animal invents machinery, or spins, or weaves, or requires or uses clothing. Wheat, corn, and the other cereals—which, if not coeval with man, were, apparently at least, the result of man's cultivation, and would disappear if man ceased to cherish them—are the staple support of the human race, and sometimes also of the domesticated animals dependent on man; but no other animal sows, or tends, or reaps. Other vegetable as well as mineral substances afford medicines for disease; but no other animal can detect the cause of disorder, can extract the drug, prescribe the remedy, or effect the cure. Others present the pigment wherewith man can paint a scene or decorate a dwelling; but no other animal uses a brush, or comprehends its power, or is gratified by its product. Nay, the very landscape in all its beauty, the sky in all its splendour, the flower in all its loveliness, have no charms for the inferior creature. To their unconsciousness they are as a sealed book. For man alone of all the creatures on Earth were this beauty, and splendour, and loveliness provided, that, made in the Creator's image, he might share the Creator's joy when He beheld what He had made, and pronounced it to be good.

But, indeed, it would be possible to show that everything on Earth is contrived to contribute directly or indirectly to the benefit of man,—that even the noxious animals and the poisonous plants and minerals have ends in this way to subserve. And if this be so, and who can doubt it, do we not find here that life on Earth has one special end, that of ministering to the rational being, Man, who, unless the Earth had been so prepared for him, could not have existed upon it, or have existed upon it as a rational being. So that, in other words, the Earth was not only specially adapted, but specially made for man,—again proving the miraculous intervention of the Supreme Being. And if so, what reason can there be for presuming life on other globes. Presumption for it thus disappears, and leaves little further to be said.

Yet the matter by no means rests here, or is altogether dependent on these views. For we must now proceed to examine the whole solar system—the only planetary system of which we, with certainty, know, or which can at present be brought, if any other exist, within our ken, and see whether what we can learn regarding the condition of the other planetary bodies warrants the assumption of resemblance to the Earth, or gives any countenance to the idea that they can be inhabited. If it do not, then a still stronger argument arises against life in other systems. For, if even the displacement of a single planet from being a life-bearing body be a fatal blow to the plurality thought, how much more conclusive will the fact be if there be found reason sufficient to displace them all.

V.

AND naturally we begin with

THE MOON.

For it is similarly distant from the Sun, and being the body nearest by far to us, we have greater facilities for its examination than we possess in regard to any other of the members of the solar system. And it is well also to keep in view its close affinity, for, upon the principle of the formation of the planetary bodies according to the Nebular hypothesis, the Moon must have been formed either out of a ring separated from the Earth,—the most probable supposition,—or from the same zone out of which the Earth was formed. Consequently, and especially if regarded as the offspring of the Earth, there ought to be a close resemblance or family likeness between the two bodies, a parity of condition far beyond what might be expected to be found subsisting between the Earth and any other planet of our system.

Sir William Herschel lent the authority of his name to the supposition that the Moon is inhabited. In a passage quoted by Sir David Brewster, who clung to his skirts in this matter, Sir William said, 'I believe the analogies that have been mentioned sufficient to establish the high probability of the Moon's being inhabited like the Earth.'[1] And Sir David sets himself

[1] *More Worlds than One*, p. 99.

assiduously to assign reasons, of which the main one was rested on imagined discovery of volcanoes in the Moon, as doubtless inferring the presence of a supporting atmosphere (though fire and smoke are not very friendly to life) for a similar opinion; but winds up by asserting that 'independent of these considerations, we maintain that every planet and satellite in the solar system *must* have an atmosphere.'[1]

The matter, however, is not one to be decided on authority, but upon fact and deduction from fact, upon observation and scientific consideration.

And the first remark which falls to be made is the extraordinary appearance which the Moon presents to view when examined by a powerful telescope. For it is studded all over with mountains rising to a great height and of peculiar shape, and rugged, bare, and terrifying. Dr. Dick,[2] who furnishes a pretty full description, classifies them into—(1) insulated mountains rising from level plains like a sugar loaf, thirty of which are from two to five miles in perpendicular height; (2) ranges of mountains extending in length two or three hundred miles, in some places four miles in perpendicular altitude; (3) circular ranges from 1/5th of a mile to $3\frac{1}{2}$ miles high, and enclosing plains from 40 to 50 miles in diameter; (4) central mountains in the midst of circular plains. There are, likewise, he mentions, numerous caverns, whose depth below the surface of the Moon varies from one-third of a mile to three miles and a half.

'Twelve of these cavities as measured by Schroeter are found to be above two miles in perpendicular depth. These cavities constitute a *peculiar* feature in the scenery

[1] *More Worlds than One*, p. 107.   [2] *Celestial Scenery*, p. 234.

of the Moon and in her physical constitution, which bears scarcely any analogy to what we observe in the physical arrangements of our globe.'[1]

Rambosson says that the circular mountainous enclosures have diameters of 130 or 140 miles, so immensely larger than the craters of our volcanic mountains as to suggest forces of terrific grandeur sadly inimical to life.[2] These are thought to be volcanic mountains; but Sir John Herschel mentions that there are here and there chains of mountains whose appearance suggests no suspicion of volcanic origin.[3]

Although Sir John also says that there are large regions perfectly level, and apparently of a decided alluvial character, it would strike one that, apart from other considerations, to be immediately adverted to, a world so formed—to say nothing as to whether it would or would not ever be adapted for the growth of plants and trees, to which water as well as air are necessary—would not offer any attractions as a residence for a being like man. It would be dismal, horrible, and unfit for intelligent beings. Nor is there, upon examination of the lunar surface, any trace visible either of vegetation or of intelligent life. There are no patches of green, such as might indicate forests or grassy fields; no cities or extensive works, to indicate the operations of rational beings.

Professor Nichol tells us that the Moon

'shines with most various colours. In the *Mare Serenitatis*, for instance, large districts shine with a most beautiful green, intermingled with grey portions and several other tints. These varieties seem constant, and therefore indicate differences on the Moon's soil or surface, suggesting the different aspects of light reflected

[1] *Celestial Scenery*, p. 245.   [2] *Astronomy*, p. 190.   [3] *Outlines*, § 430.

NO APPEARANCE OF LIFE. 221

from a desert of Sahara, a golden savannah in the New World, or a field of luxuriant and freshening fern.'[1]

But this statement requires to be taken along with that made by Mr. Proctor, that 'although there are varieties, *there has never yet been detected any variation, of colour.* Nothing has been seen which could be ascribed with any appearance of probability to the effects of seasonal change.'[2]

So that vegetation evidently does not exist. And as for changes resulting from the action of beings like man, Mr. Proctor—stating that we could only see edifices far greater than those yet constructed by man (which is surely much overstating the case), but that from the less force of gravity at the Moon's surface it would both be needful and in the power of inhabitants to build to much greater size than we do—concludes by saying, that though such structures might be visible in our largest telescopes, '*no object that could with the slightest appearance of probability be ascribed to the labours of intelligent creatures, has ever been detected on the Moon's surface.*'

Dr. Whewell,[3] mentioning that Professor Phillips had stated, at the meeting of the British Association in 1853, that astronomers could discern the shape of a spot on the Moon's surface of a few hundred feet in breadth, and that such changes as have occurred upon the Earth would be visible through telescopes in the Moon, remarks that we should see like changes, if they are going on, upon the face of the Moon.

'Yet no such changes,' he says, 'have ever been noticed. Nor even have such changes been remarked as might occur in a mere brute mass without life,— the formation of new streams of lava, new craters, new

---
[1] *Phenomena of Solar System*, p. 170.   [2] Proctor's *Moon*, p. 263.
[3] *Of the Plurality of Worlds: an Essay*, p. 289.

crevices, new elevations. The Moon exhibits strong evidences, which strike all telescopic observers, of an action resembling, in many respects, volcanic action, by which its present surface has been formed. But if it have been produced by such internal fires, the fires seem to be extinguished, the volcanoes to be burnt out. It is a mere cinder, a collection of sheets of rigid slag and inactive craters. And if the Moon and the Earth were both, at first, in a condition in which igneous irruptions from their interior produced the ridges and cones which roughen their surfaces, *the Earth has had this state succeeded by a series of states of life in innumerable forms,* till at last it has become the dwelling-place of man; *while the Moon, smaller in dimensions, has at an earlier period completely cooled down, as to its exterior at least, without ever being judged fit or worthy by its Creator of being the seat of life,* and remains hung in the sky as an object on which man may gaze, and perhaps from which he may learn something of the constitution of the Universe; and among other lessons this: that he must not take for granted that all the other globes of the solar system are tenanted, like that on which he has his appointed place.'

Still we have been dealing hitherto only with negative considerations, we must now look to the reasons which exist for a positive exclusion of life.

And the first of these is the absence in the Moon of the slightest trace of water or of cloud. Upon this all astronomers are agreed.

Thus Professor Nichol, referring to a map of the Moon prepared by Messrs. Baer & Mädler of Berlin, says:—

'On the face of that map there may be seen a number of comparatively dark and untroubled spots, visible in general to the naked eye, and which were at first thought to be *seas*, whence their present names. Not a drop of water, however, or of any fluid akin to it, is now in the Moon; for when the line of light in the crescent luminary passes through these so-called seas, its edge, instead of following a fine and unbroken course, is quite irregular, proving that those flats are *land*, although comparatively unbroken and undisturbed—probably they are in some cases the *subsided* parts of the outer shell or bed of that globe. Besides, there are no clouds; and as the Moon has a small atmosphere, this alone is decisive; for otherwise vapours would float there, and the purity of the disc appear variable.'[1]

Dr. Dick[2] also observes that the so thought seas are simply extensive plains diversified with gentle elevations and depressions, and consisting of substances calculated to reflect the light of the Sun with *a less degree of intensity* than the other parts of the lunar surface.

Professor George Wilson[3] says: 'The Moon has no clouds or rain, nor dew, nor lakes, nor rivers, nor seas.'

Mr. Proctor says:—

'There are no seas or oceans in the Moon. Were there any large tracts of water, the tremendous heat to which the Moon is subjected during the course of the long lunar day (lasting a fortnight of our time) would certainly cause enormous quantities of water to evaporate; and not only would the effects of this process be distinctly recognisable by our telescopists, but the spectroscope would exhibit in an unmistakeable manner the presence of the aqueous vapour thus formed.'[4]

A supposition has indeed been made, that at one time the Moon possessed seas, but that the water has been drawn to the interior by the force of gravity or of pressure, and that the same process is proceeding, or has

---

[1] *Phenomena*, p. 161.
[2] *Celestial Scenery*, p. 248.
[3] *Chemistry of the Stars*, p. 25.
[4] *Other Worlds than Ours*, p. 176.

proceeded, upon the Earth and other planets. But I am not aware that this theory rests on any reliable evidence or data; and it is opposed by many objections, the more especially if it be assumed as necessary to its maintenance that the Moon is older than the Earth, and the interior planets are older than the exterior, which would seem to be in opposition to the views obtained from the Nebular hypothesis. Supposing, however, that water had at one time existed and was now so withdrawn from the Moon, would not the influence of the Earth's attraction, exerted upon it with a force more than tenfold the power which the Moon exerts upon our seas, together with the, to us, horrible nature of the Moon's surface, have rendered our satellite entirely unfit for the abode of rational beings? The likelihood, indeed, rather is, that the Moon, as the smaller body, cooled down while the Earth was still in its original nebular glow, and perhaps extended far out towards the Moon itself, and, it may be, attracted to it every particle of moisture which could form into the shape of water. Leaving this theory, which remains to be substantiated, —and if true, would not necessarily infer life, life of any kind while the fluid lasted,—it is clear that no seas do exist in the Moon.[1]

Equally important is it that there does not exist in the Moon an atmosphere. On this philosophers are all but agreed; and even those who suppose that any atmosphere exists, believe it to be very scanty and finely attenuated. Nor does it at all follow that if there were any, it would be life-supporting. This is a

[1] Supposing this theory maintainable as regards the other planets, it would not suffice to have made them habitable, and does not meet the positions with regard to them which have been already and will be stated.

circumstance never sufficiently regarded, and which I cannot press too strongly. However, there seems no room to hold that an atmosphere of any kind encircles the Moon.

At one time the luminous corona of the Sun, seen during an eclipse to surround the Moon, was believed to be due to a lunar atmosphere. But this idea has long since been disproved and abandoned.

Sir John Herschel says, that were there any atmosphere it could not fail to be perceived in the occultations of stars and the phenomena of solar eclipses, as well as a great variety of other phenomena; and concludes, 'the non-existence of any atmosphere at the Moon's edge dense enough to cause a refraction of 1″, *i.e.* having 1/1980th part of the density of the Earth's atmosphere.'[1]

Mr. Grant, in his *History of Physical Astronomy*,[2] discusses the subject carefully.

From the uniform appearance presented by the lunar spots, he says, it is reasonably supposed that they are not occasionally obscured by clouds, and hence the Moon cannot be, like the Earth, encompassed by a gaseous fluid.

Again, he points to the fact of there being no sensible change of colour and diminution of lustre on stars passing into occultation as another evidence; and the very rare cases in which the contrary has been observed not being general, cannot establish the existence of a lunar atmosphere. In like manner he disposes of individual cases where, when a star has come up to the Moon, it has been observed to hang on the margin of its disc for three or four seconds previous to disappearance, a circumstance sufficiently disposed of by the observation of Mr. Proctor, after mentioned.[3]

[1] Herschel's *Outlines*, § 431.   [2] Grant, p. 230.   [3] Proctor's *Moon*, p. 290

Another reason is, that were there an atmosphere, its refraction would cause the star to be visible for a short time previous to disappearance, and for a like period prior to its emergence; but calculation shows there is no such refraction.

He adduces in further evidence of there being no atmosphere to cause refraction,[1] the circumstance that planets do not, in coming in contact with the limb of the Moon, suffer distortion of figure. He also refers to certain investigations of Du Sejour, a French astronomer, upon supposed refraction exhibited during solar eclipses, which, if well founded, would give a lunar atmosphere 1400 times rarer than common atmospheric air, and consequently 'exceeding the most perfect vacuum which had been hitherto formed by means of the air-pump.'

Lastly, astronomers had failed to discover any twilight, though Schroeter had at length succeeded in observing such an appearance, which, if correct, would give a small amount of atmosphere. It would appear, however, that Schroeter must, for the reasons assigned by Mr. Proctor,[2] have been mistaken, and that in reality there is no twilight, and hence no atmosphere.

Mr. Proctor, in his work on the Moon,[3] shows that, supposing the Moon had an atmosphere equal to our own, it would, under the small attractive power of the body, extend to a height of 22 miles, and would be so attenuated that 'men would perish if placed' in it.

It will be readily seen that if, instead of being equal to our own, it is so minute in quantity, as is supposed by these philosophers, life would be altogether impossible.

A heavier gas, Mr. Proctor proceeds to show, would produce the effect of refraction, and even were the atmo-

---

[1] See Leitch's *God's Glory in the Heavens*, pp. 38, 39.
[2] Proctor's *Moon*, p. 297.   [3] *Ibid.* pp. 286, 287.

sphere possessed of only a fourth part of the refractive power of our own, a star in reality behind the Moon's disc would appear as a ring of light, the light from the star being increased as by an enormous lens.

Mr. Proctor[1] further states that sometimes a star has been observed to diminish just before disappearance, which may be attributable to the star being a close binary or a multiple star, or perhaps a nebulous star; and that, on the other hand, a star has seemed to advance for some distance on the Moon's disc before vanishing, which may be attributed to its chancing[2] 'to cross the Moon's limb where a valley or ravine has caused a notch or depression which is too small to be indicated by any ordinary method of observation.' He adds that the evidence thus obtained is strengthened by spectroscopic evidence, Dr. Huggins finding that the spectrum of a star disappears as instantaneously as the star itself.

But the strongest evidence, in Mr. Proctor's view,[3] that the Moon has no atmosphere, or so little that she may be regarded as practically airless, is that in a solar eclipse the Sun would not disappear at all, even at the moment of central eclipse, because if there were a lunar atmosphere it would act as a lens, and reveal the Sun as a ring of blazing lustre, 'as really sunlight as the light of our setting Sun,'—'a very shallow lunar atmosphere would suffice.'

It was at one time thought, as already mentioned, that the existence of active volcanoes had been discovered in the Moon, and that this, as indicating fire, would prove an atmosphere.

'As soon,' says Professor Leitch,[4] 'as the telescope unveiled a world so like our own in its general aspect,

[1] Proctor's *Moon*, p. 288. [2] *Ibid.* p. 290.
[3] *Ibid.* p. 293. [4] *Ibid.* p. 38.

the popular imagination peopled it with living forms, and astronomers strove with one another to gratify the popular wish. On the dark surface active volcanoes were discovered, the flames were seen to burst forth with great fierceness, and then slowly expire, showing that there was air to sustain the combustion. Planets and fixed stars were seen to linger on the Moon's edge before they passed behind its disc, just as they ought do if there was an atmosphere. All these observations are now discredited; and the inexorable decision of scientific research is that there is no valid argument for the existence of an atmosphere.'

That astronomers, however, and men like Sir William Herschel, have fancied they had perceived volcanoes is true; and Sir David Brewster, observing, in his usual positive manner,[1] 'that volcanoes or burning regions have been observed in the dark portion of the Moon's disc *cannot be doubted*,' adduces the supposed instances.

But Schroeter and others, even including Dr. Dick[2] (prone as he is, like Sir David Brewster, to contend for the habitability of the orbs), ascribe the appearances to light reflected from the earth.

Mr. Proctor in his *Moon* discusses the supposed instances, showing how unreliable were the observations,[3] although changes such as have been supposed might probably arise from other causes than fire, and concludes, 'the inference would seem to be that the supposed changes have been *merely optical*.'

Perhaps the most conclusive evidence of all, however, is afforded by spectrum analysis.

Mr. Proctor[4] alludes to this, remarking that 'when the spectrum of the lunar light has been observed (by Dr. Huggins first, and later by others) it has

[1] Brewster, p. 105.
[2] Dick's *Celestial Scenery*, pp. 246, 247; Proctor's *Moon*, p. 264.
[3] Proctor's *Moon*, p. 272.   [4] *Ibid.* p. 292.

been found to be absolutely similar to the solar spectrum, that is, there is no trace whatever of absorptive action exerted by a lunar atmosphere upon the solar rays which are reflected by her to the Earth. This evidence is, of course, not demonstrative of the absolute want of air of any sort in the Moon, because a very rare and shallow atmosphere would produce no appreciable absorptive effect; but it confirms the other evidence, showing that any lunar atmosphere must not only be extremely shallow, but extremely rare.' 'An atmosphere sufficient in quantity to give traces of its presence in lunar shallows, but not extending higher than the summits of the lunar mountains, must be of a specific gravity so greatly exceeding (under the same conditions) that of common air, or indeed of any gas known to us on earth, that we are justified in regarding the theory of its existence as altogether unsupported by evidence.'

'The conclusion to which we seem forced by all the evidence obtainable is,[1] that either the Moon has no atmosphere at all (which seems scarcely possible), or that her atmosphere is of such extreme tenuity as not to be perceptible by any means of observation we can apply.'

This slender atmosphere, if it exist, is not, as we have already seen, capable of sustaining life. But the spectroscopists state the same thing, and even more unqualifiedly. Mr. Roscoe, referring to the observations of Dr. Huggins and Dr. Miller, says :—[2]

'The conclusion must be drawn that the Moon is *devoid of any appreciable atmosphere.*'

And Schellen—[3]

'The observations of Fraunhofer (1823), Brewster, and Gladstone (1860), Huggins and Miller, as well as Janssen, agree in establishing the complete accordance of the lunar spectrum with that of the Sun. In all the various portions of the Moon's disc brought under observation no

[1] Proctor's *Moon*, p. 298.   [2] Roscoe, p. 230.   [3] Schellen, p. 481.

difference could be perceived on the dark lines of the spectrum, either in respect of their number or relative intensity. From this entire absence of any special absorption lines it must be concluded that there is no atmosphere in the Moon, a conclusion previously arrived at from the circumstance that during an occultation no refraction is perceived on the Moon's limb when a star disappears behind the disc. Moreover, a small telescope of only a few inches' aperture suffices to show the spectrum of the Moon very distinctly.'

But what for sometime was regarded as a grand discovery, calculated to upset the notion that the Moon is entirely uninhabitable, was made or thought to be made by M. Hansen. Professor Leitch, introducing the matter with a great flourish of trumpets, tells us that the cause of an irregularity in the Moon's path was solved by M. Hansen finding the centre of gravity and the centre of the figure of the Moon to be

'distant about 37 miles from the other. Most momentous results flow from this. The one hemisphere must be lighter than the other.'[1] The application is very direct and startling. 'Supposing the sphere of the Moon originally covered with water, and enveloped in an atmosphere, both water and air would flow to the heavier side, and leave the lighter side destitute of both, just as water and air leave the summits of our mountains, and gravitate towards the valleys. They seek the lowest level, or in other words, the point least distant from the centre of gravity. In the case of the Moon, the side turned to us is virtually one enormous mountain, and the opposite side the corresponding valley. We could not expect to find traces of air on the summit of a terrestrial mountain 134 miles high. The conclusion, therefore, is that though the near hemisphere is a lifeless desert, having neither water nor air to sustain life, the hidden

[1] *God's Glory in the Heavens*, p. 42.

hemisphere *may* have a teeming population, rejoicing in all the comforts and amenities of life. The imagination is set free to picture broad oceans, bearing on their bosom the commerce of this new world, rivers fertilizing the valleys through which they flow, a luxuriant vegetation, and buildings of colossal size.'[1]

Mr. Leitch devotes a whole chapter to this discovery. Sir John Herschel[2] adopted it as not improbable, and it was currently credited by others. One would have thought that the attraction of the Earth would rather have drawn away air and water on the Moon towards the side exposed to the influence, just as the Moon, with a greatly less power, influences tides upon the Earth, and, failing water and air, that the attraction would be as suggested upon the body of the planet itself, causing it to bulge.

The question has, however, been taken up by Mr. Proctor with his usual ability, and, referring to examinations made by M. Gussew[3] and others, he says :—

'But pleasing though the idea may be that on the farther hemisphere of the Moon there may be oceans and an atmosphere, *it appears to me impossible to accept this theory.* In the first place, it has not been demonstrated, and is in fact not in accordance with theoretical considerations, that the Moon is egg-shaped, or bispherical, according to Gussew's view. The farther part may also project, as the nearer part does (supposing Gussew's measurements and inferences to be trustworthy). But even if we assume the Moon to have the figure assigned to it by Gussew, the invisible part is not that towards which the atmosphere would tend. The part of the surface opposite the centre of the visible disc is in fact not nearest to the centre of gravity, but (assuming the

[1] *God's Glory in the Heavens*, p. 43.
[2] Herschel's *Outlines*, § 436, *a* and *b*.   [3] Proctor's *Moon*, p. 301.

unseen part spherical, and of the radius indicated by the visible disc) is 30 miles farther from the centre of gravity than are points on the edge of the visible disc. The band or zone of the Moon's surface lying on this edge is the region where oceans and an atmosphere should be collected (if water and air existed in appreciable quantity) on the Moon's surface.'

In a footnote he quotes the argument presented in another form, as contained in a paper contributed by him to the monthly notices of the Astronomical Society.

In a later work (*Expanse of Heaven*, p. 45) Mr. Proctor gives an additional reason against the existence of water and air on the farther side of the Moon, in the fact that we see partly round it—the parts seen at one time amounting in all to 589 out of 1000, the parts never seen amounting to only 411, and no signs of air or ocean detected on those revealed portions.

Mr. Kalley Miller,[1] giving also a lively account of the supposed discovery resulting from the Moon being observed on a certain occasion to have got no less than three seconds behind her proper calculation, assigns the shattering of the theory to Professor Adams, who

'showed that the calculated three seconds' discrepancy could be accounted for by more natural causes,' proving 'that the Moon's eccentricity was both unnecessary and untenable, and dealt a death-blow to the antipodean Selenites, by cutting off their water supply and putting them under an air pump.'

Dismissing, therefore, the idea of an atmosphere on either hemisphere of the Moon, some strange results follow, and are thus described by Flammarion:—[2]

---

[1] *Romance of Astronomy*, pp. 80-85.
[2] Flammarion's *Marvels of the Heavens*, p. 246.

'On approaching the Moon nothing is seen of the physical causes which make the Earth a vast laboratory wherein a thousand elements contend or unite with each other. There are none of those tumultuous tempests which sometimes sweep over our inundated plains; none of those hurricanes which descend in waterspouts, to be swallowed up in the depths of the sea; no wind blows, no cloud rises to the heavens. There white trains of cloudy vapours are not seen, nor those leaden masses with heavy cohorts; the rain never falls; and neither snow nor hail nor any of the meteorological phenomena are manifested there.

'But, on the other hand, the magnificent tints which colour our sky at sunrise and twilight, the radiations of the heated atmosphere, are never seen there; if winds and tempests never blow, neither is there the balmy breeze which descends upon our coasts. In this kingdom of sovereign immobility, the lightest zephyr never comes to caress the hill tops; the sky remains eternally asleep in a calm incomparably more complete than that of our hottest days, when not a leaf moves in the air. This is because on the surface of this strange world there is no atmosphere. From this privation results a state of things difficult to realize. In the first place, the absence of air implies also the absence of water and every liquid, for water and liquids can only exist under atmospheric pressure; if this pressure is taken away they evaporate, and their beds are dried up. ... From the absence of air follows another very curious fact—the absence of sky. On the surface of the Moon, when the looks are directed towards the sky, there is none to be seen. An immensity without depth is traversed by the sight without resting on any kind of form, and in the day as in the night are seen the stars, planets, comets, and all the bodies of our Universe. The Sun passes among them without extinguishing them, as it does to us. Not only does the Moon not possess this perpetual diversity which the movements of the air produce in our world, but it has not the azure vault which covers the Earth with such a magnificent dome; space is a black, and a perpetually black, abyss.

'Whilst on high there reigns darkness, below there is silence. Not the least sound is ever heard; the sigh of the wind in the woods, the rustling of foliage, the song of the morning lark, or the sweet warbling of the nightingale never awakens the eternally dumb echoes of this world. No voice, no speech has ever disturbed the intense solitude with which it is overspread. Unchangeable silence reigns there in sovereignty.'

Similarly Professor Leitch says:—

'Eternal silence must reign there. A huge rock may be precipitated from the lofty lunar cliffs, but no sound is heard—it falls noiselessly as a flock of wool. The inhabitants can converse only by signs. The musician in vain attempts to elicit sweet music from his stringed instrument; no note ever reaches the ear. Armies in battle array do not hear the boom of the cannon, though rifled arms, from the low trajectory of the ball, must acquire a fatal precision and range. No moving thing can live aloft; the eagle flaps its wings against the rocks, and in vain attempts to rise; the balloon instead of raising the car, crushes it with the weight of its imprisoned gas.'[1]

These are the descriptions of two advocates of plurality. It is evident that even were life possible in the Moon, it could not be the abode in such circumstances of intelligent beings.

But there is a still further fact now to be stated—a fact destructive of any life, and therefore, and in any view taken along with the absence of water and air, conclusive against the habitability of the Moon. It lies in the enormous length of the lunar days and nights. In the words of Mr. Proctor:—

'The lunar day lasts about a fortnight, and the lunar

[1] *God's Glory in the Heavens*, p. 33.

night is, of course, equally long. Were this all, the inconvenience of the arrangement would be unbearable by beings like ourselves. But far more serious consequences must result from the combination of the arrangement with the want of an atmosphere; for whereas during the lunar day the surface of the Moon is exposed to an inconceivably intense direct heat, undoubtedly sufficient to heat that surface far above the boiling point, during the lunar night the heat is radiated rapidly away into space (no atmosphere checking the process), and an intensity of cold must prevail of which we can form but imperfect conceptions.'[1]

With this agree the statements of Rambosson, p. 191; Herschel, §§ 431, 432; Leitch, p. 34, and others. Professor Leitch says:—

'The inhabitants having no atmosphere to shelter them from the Sun, and store up its heat, must recoil with terror from its fierce rays. During the long lunar day the ground must become as burning marl, from which the scorched feet shrink with pain; during the equally long night it must be colder than frozen mercury. No fuel will burn to mitigate the rigour of the cold, and none but the electric light can avail to dispel the darkness.'

'It has been calculated,' says Cooley, 'that the side of the Moon facing the Earth has a few days after full Moon a maximum temperature of 750° F. (400° C.), and again soon after new Moon sinks to $-187°$ F. ($-122°$ C.).'[2] Lord Rosse was enabled to estimate the difference of temperature to be more than 500°, so that probably when the Sun was not shining on it the temperature would be two or three hundred degrees below zero, which, under the vertical Sun, it

---

[1] *Other Worlds than Ours*, p. 177.
[2] Cooley's *Physical Geography*, p. 106.

would be as much above zero, or hotter than boiling water.[1]

Vegetation and life of all kinds could not abide exposure to such scorching heat or to such inconceivable cold. The sudden and constant alternation from a far less extent of either would also of itself be equally fatal. We are therefore shut up to the conclusion, that the Moon—a body so large as to contain a surface of 14,600,000 square miles—is incapable of life, and in this opinion all, or nearly all, are now agreed.

But the fact has a most telling bearing upon the present inquiry. It is not the mere displacement of one solitary body from the world of life; it breaks the argument for all.

Among the writers on plurality I have only noticed one—Dr. Whewell—who has perceived the deduction to which this momentous fact inevitably leads. He thus forcibly puts it:—

'If it appear by strong evidence that the Moon is not inhabited, then is there an end of the general principle that *all* the bodies of the solar system are inhabited, and that we must begin our speculations about each with this assumption. If the Moon be not inhabited, then, it would seem, the belief that each special body in the system is inhabited must depend upon reasons specially belonging to that body, and cannot be taken for granted without such reasons. Of the two bodies of the solar system which alone we can examine closely, so as to know anything about them, the Earth and the Moon, if the one be inhabited and the other blank of inhabitants, we have then no right to assume at once that any other body in the solar system belongs to the former of these classes rather than to the latter. If even under terrestrial conditions of light and heat we have a total absence of the

[1] Newcomb, p. 319.

phenomenon of life known to us only as a terrestrial phenomenon, we are surely not entitled to assume that when these conditions fail we have still the phenomenon life. We are not entitled to *assume* it; however, it may be capable of being afterwards proved in any special case by special reasons; a question afterwards to be discussed.'[1]

At the time Dr. Whewell wrote, it will be kept in view further that spectrum analysis had not been employed to the extent it has now been in the investigation of the stars.

[1] *Essay*, p. 288.

VI.

But leaving the Moon, we shall proceed to consider the various planets, commencing with the minor ones, and with Mercury, the nearest to the Sun; for I put aside Vulcan, which, notwithstanding the circumstantial account of its supposed discovery by the rustic French doctor, as graphically narrated by Professor Leitch,[1] is now generally discarded as having been an illusion; or, at least, it is regarded as of very doubtful reality.[2] Had it really existed, the blaze of heat alone to which its proximity to the Sun would subject it would have scorched up and destroyed every species of life upon its surface.

MERCURY.

This small planet is so comparatively near to the Sun that it is not well adapted for observation, and on various points opinions have in consequence differed, but most of these points are not very material to the present inquiry. Its year is found to be 88 days, and its day is about 24 hours, or slightly more. Its velocity in its orbit is 109,360 miles per hour, and it has no satellites. Upon these points, and also upon its distance from the Sun, there is a general agreement. But in regard to its size, density, mass, weight, surface, gravity, there is more or less difference. It has been said to have lofty mountains, ten or twelve miles high.[3] The

---

[1] *God's Glory in the Heavens*, p. 81 et seq.   [2] Lardner, § 309.
Rambosson, p. 113; Grant, p. 233.

points upon which it is here material to dwell are the inclination of its axis, its varying distance from the Sun, its light and heat, and the question of its atmosphere.

It has been stated that the axis is 7 degrees 'inclined to the plane of the equator, so that there must be a great inequality in the days and seasons.'[1] But although it is generally considered that there is such an inclination, the point is one which does not appear to be conclusively settled. The safer course is to put it aside till it be so.

Upon another point there is not uncertainty. 'The orbit of Mercury,' says Sir John Herschel, 'is very elliptical, the eccentricity being nearly one-fourth of the mean distance,'[2] which may be stated to be in round numbers 36,000,000 miles. Lardner states (others agreeing) the distance from the Sun in aphelion to be $42\frac{1}{2}$ millions, and in perihelion to be 28 millions of miles. The practical effect of this amazing difference in the Mercurial year of 88 days is mentioned by Mr. Proctor :—

'When he is nearest to the Sun he receives *ten and a half times more light and heat* from that luminary *than we do;* but when he removes to his greatest distance, the light and heat he receives are reduced by more than one half. Even then, however,' he adds, 'the sun blazes in the skies of Mercury with a disc four and a half times larger than that which he presents to the observer on Earth.'[3]

Fontenelle himself seems, from a consideration of the

---

[1] Rambosson, p. 115.   [2] *Outlines*, § 482.
[3] *More Worlds than Ours*, p. 58.

tremendous heat, almost to have renounced the thought of its habitability.

'But what do you think, then, of the inhabitants of Mercury. They are yet nearer to the Sun, and are so full of fire that they are absolutely mad. I fancy they have not any memory at all, no more than most of the negroes; that they make no reflections, and what they do is by sudden starts, and perfect haphazard; in short, Mercury is the Bedlam of the Universe. The Sun appears to them much greater than it does to us, because they are much nearer to it than we; it sends them so vast and strong a light that the most glorious day here would be no more with them than a declining twilight. I know not whether they can distinguish objects, but the heat to which they are accustomed is so excessive that they would be starved with cold in the torrid zone.'[1]

And in order to relieve the supposed inhabitants he says:—

'In China there are countries which are extremely hot by their situation, yet in July and August are so cold that the rivers are frozen. The reason is, they are full of saltpetre, which, being exhaled in great abundance by the excessive heat of the Sun, makes a perfect winter at midsummer. We will fill the little planet with saltpetre, and let the Sun shine as hot as he pleases.'[2]

Very different is the account given by Flammarion as the view of one of the numerous romancers who feigned voyages o the planets, and who

'pretended to know that the mountains of Mercury were all crowned with beautiful gardens, in which grew naturally not only the most succulent fruits, which served as food to the Mercurians, but also the greatest variety of dishes. It would appear that in this happy world it is not necessary to prepare, as with us, things for food. Fowls, hams, beefsteaks, cutlets, entremets,

[1] Fontenelle, p. 99. [2] *Ibid.* p. 101.

small side dishes, etc., were raised there in the same manner as the apples on our apple trees, and when a repast was wished for it was sufficient to spread the cloth; then arrived bird-waiters for your orders, who knowingly flew away, and in the twinkling of an eye, from the mountains, where the desired dishes are to be found, brought you them with the greatest haste. It is better, perhaps, to believe that the vegetables of Mercury possess these precious gifts, and that its birds are of such delightful intelligence, than to think, with Fontenelle, that the inhabitants of Mercury are all mad, and that their brains are burned with the violent heat which the Sun pours upon their heads.'[1]

To avoid this enormous heat, Mr. Proctor suggests that possibly if the planet's axis be not so much inclined as it is considered to be, the polar regions 'may not form a disagreeable abode;'[2] but he further suggests the probability of an atmosphere of great rarity,[3] and that from the analogy of lofty mountains on the Earth, which are cooler than the plains below, yet goes on to say:—

'We cannot thin the Mercurial air without adding to the direct effects of the Sun upon the Mercurial inhabitants. Whether in this way we increase the habitability of the planet may be doubted when we consider that the direct action of the Sun's rays upon the tropical regions of Mercury, thus deprived of atmospheric protection, would produce a heat four or five times greater than that of boiling water.'[4]

Dissatisfied with this theory, he proceeds to take up the opposite one, of there being a dense atmosphere surrounding Mercury.[5] And this is a view which others have advocated to mitigate the consuming heat. But

[1] Flammarion's *Marvels*, p. 160.
[2] *Other Worlds than Ours*, p. 61.
[3] *Ibid.* p. 62.
[4] *Ibid.* p. 64.
[5] See Guillemin's *Heavens*, p. 62.

this theory he mentions is not satisfactorily established,[1] and is disproved by the fact that the planet does not, as it would were it surrounded by a cloudy envelope, shine more brilliantly. Indeed it does not, although so favourably situated for doing so, reflect the same proportion of light as some of the other planets do.

Driven from this position, and also from supposing it to be surrounded by a double cloud envelope, he thinks the reader may probably prefer the thought of a polar residence with tunnels of communication.

All this theorizing, and along with it speculation regarding the nature of the atmosphere and its life sustaining power, has been put at rest by the observations made a few years ago by Dr. Zöllner by the use of his photometer:—

'This instrument is a very beautiful application of the principle that if light be reflected from a smooth surface, such as a plate of glass inclined at a certain angle, it undergoes a modification of such a nature that the amount reflected from a second plate inclined to it at the same angle varies as the plate is turned about the ray as an axis, being greatest when the two plates are parallel, and nothing when the second plate is turned round through a right angle from this position.'[2] 'From his measures of the brightness of Mercury with this photometer, he concludes that the reflective power of this planet is much the same as that of the Moon (being a little higher than quartz porphyry), and that the change of brightness with the phase is just what would hold for a rough surface in which the average slope of the hills was 52°; *from this he infers that Mercury must therefore be without an atmosphere.* The reflective power of the other planets, with the exception of Mars, is very

[1] *Other Worlds than Ours*, p. 66.
[2] *Year Book of Facts in Science and Arts for* 1875, p. 251.

much larger, and they have therefore probably very dense atmospheres, which reflect nearly all the Sun's rays.'

It is therefore obvious that Mercury cannot be inhabited. But further, as it has no atmosphere and no clouds, it can have no water; and, indeed, the great heat of the Sun would speedily cause all water to evaporate and disappear from the planet. There being no water, there can be no vegetation to support life—a fact which those anxious to secure inhabitants for the planets seem regularly to ignore, or rather to forget.

We thus start with first depopulating the Moon, and then the planet nearest to the Sun. Let us see whether we find any different state of matters when we come to the second planet.

VENUS.

In some things this planet resembles the Earth. It is very nearly the same in size (the diameter being a little over 200 miles less), in density volume and mass, gravity, and length of day. These similarities have been founded on as inferring the habitability of Venus. But, in reality, they do not affect the question, which must be determined by other circumstances. It should, however, be mentioned that there are circumstances, not material to the present inquiry, in which Venus differs from the Earth. Various observers, from time to time, fancied they descried a satellite. But it is now considered, as stated by Lardner, 'that the supposed appearances were illusive.'[1] The subject is also discussed in Dick's *Celestial Scenery*, p. 80; Proctor's *Other Worlds than Ours*, p. 71; Proctor's *Orbs around us*, p. 82. Although there has not been conclusive agreement on the subject of the mountains in Venus, yet it seems generally held, and the appearances seem to justify the opinion, that the mountains in Venus are of great height—five times higher than the highest mountains on Earth; a fact which must necessarily create a considerable difference on the natural features of the planet, and may indicate at one time an appalling amount of volcanic energy, which possibly may have pervaded the whole surface, and corrugated it to such an extent as to render it unfit for cultivation and habitation. See Rambosson, p. 122; Herschel's *Out-*

[1] Lardner's *Handbook of Astronomy*, § 345.

lines, § 509; Dick's *Celestial Scenery*, pp. 77, 78; Grant's *History*, p. 235; Proctor's *Orbs around us*, p. 90; Nichol's *Phenomena*, p. 150.

But although mountains apparently exist, the features of the planet are indiscernible, and it is not disputed that there is no appearance of seas. The only things seen are spots, and the spots have not always been perceptible — long periods of time having elapsed between observations of them. See Dick, pp. 73, 74; Grant, pp. 234–236; Proctor's *Orbs*, p. 91. By watching these spots, however, astronomers have been enabled to calculate the period of its rotation on its axis, or length of its day.

But the planet is not well situated for observation, principally because when nearest to the Earth we see the side which is in shade, and because when in opposition the surface is of dazzling lustre.

Venus is nearly 30 millions of miles nearer the Sun than is the Earth, and the light and heat showered down upon it from the Sun are about twice what are received by the Earth. Dr. Dick says:—

'It has been calculated that the greatest heat of Venus exceeds the heat of St. Thomas on the coast of Guinea or of Sumatra about as much as the heat in these places exceeds that of the Orkney Islands or that of the city of Stockholm.'[1]

Mr. Proctor says:—

'I suppose no one doubts that quite possibly this great light and heat may be so tempered as to be not only endurable, but pleasant to people in Venus. But so far as terrestrial experience is concerned, we are assuredly not justified in saying that this *must* be so. Undoubtedly, if the Sun began suddenly to pour twice as

[1] Dick's *Celestial Scenery*, p. 88.

much light and heat upon the Earth as he actually does, the human race would be destroyed in a very few months. In tropical regions the destruction would be completed in a single day. In temperate regions the beginning of the first summer would be fatal. Nor would the denizens of arctic and sub-arctic regions live through the heat of a midsummer's nightless day.'[1]

To this has to be added the fact that the orbit of Venus has little eccentricity, the difference between its perihelion and its aphelion being only one million of miles,[2] so that it may be said there can be no perceptible difference in the power or intensity of the Sun's rays throughout its whole year of 225 days.

To escape the difficulty thus arising, believers in the plurality of worlds have imagined that the intense heat may be modified by the existence of a dense atmosphere. But upon this Mr. Proctor proceeds to observe:—

'All that we know of the effects of a deep atmosphere would lead us to believe that the heat in Venus must be intensified by the action of her deep and dense atmosphere. As matter of fact it may not be so. All I urge is, that judging from the only analogy we have to guide us, the depth and density of the atmosphere of Venus seem to promise no relief from the intense solar heat to which she is exposed.'[3]

But now let us see what is said in regard to this atmosphere.

'From several of Mr. Schroeter's observations,' says Mr. Dick, 'he concludes that Venus has an atmosphere of considerable extent,' and refers to an observation made by him on 10th September 1791:—

'From these and various other observations, which it would be too tedious to detail, he concludes, on the

---

[1] *Orbs around us, Planet of Love*, p. 98.
[2] Lardner, § 325.   [3] *Orbs*, p. 99.

ground of various calculations, that the *dense* part of the atmosphere of Venus is about 16,020 feet, or somewhat above three miles high; that it must rise far above the highest mountains; that it is more opaque than that of the Moon; and that its density is a sufficient reason why we do not discover on the surface of Venus those superficial shades and varieties of appearance which are to be seen on the other planets.'[1]

Grant also says :—

'Various circumstances concur to prove that Venus is surrounded by an atmosphere of considerable density. During the transits of the planet over the Sun's disc in 1761 and 1769, the planet was observed by several astronomers to be surrounded by a faint ring of light similar to the appearance which would be occasioned by the passage of the solar rays through a circumambient fluid.'[2]

A similar ring of light was observed at the transit on 8th December 1874.[3]

Indeed, there seems general agreement regarding the existence of this atmosphere and its density, which is also proved by the circumstance that while Mercury, wanting an atmosphere, gives out little light, Venus, farther from the Sun, sheds a brilliant light.

But I think the fact goes a very great deal further than is generally supposed. Spectrum analysis appears to show that the atmosphere of Venus contains aqueous vapour,[4] and it appears to me that the excessive extent and density of the atmosphere is due to the heat to which it is subjected,[5] and that in all probability the whole seas and oceans which would otherwise exist are

---

[1] *Celestial Scenery*, p. 78.   [2] *History of Physical Astronomy*, p. 235.
[3] *Year Book of Facts in Science and Arts for* 1875, p. 252.
[4] Schellen's *Spectrum Analysis*, p. 482.
[5] Proctor's *Other Worlds than Ours*, p. 80, referring to Father Secchi's observations.

evaporated and held in moisture over it,—a supposition the more probable when it is considered that the dense portion of it extends to over three miles high, while the atmosphere itself extends, rarer, it is true, to a height of probably thirty miles. The effect of this would be to drain the oceans and to stop the rivers, and as no changes are discerned on the face of the planet there is doubtless an absence of winds to dislodge the clouds and open up the land to the rays of the Sun. Under this constant dense and probably never displaced cloud, obscuring the heavens, it is clear that no vegetation could spring up, no life could subsist, and obviously it would be an unfit abode, not merely for rational beings, but for beings of any sort.

Mr. Proctor[1] has adduced a circumstance which may prove to be, if possible, still more conclusive, in the supposed inclination of the planet's axis to the plane of her orbit. Assuming that the observation is correct, that the axis is inclined 15°, he shows that this would be fatal to human life. I would refer to what he has said on the subject. But as he adds—

'Fortunately for our belief in the habitability of Venus, astronomers are far from accepting with confidence the assertions of those observers who have assigned to Venus an inclination so remarkable.'[2]

The point may well be left until the fact be more conclusively ascertained. If the fact should really be as has been supposed,—and the difficulty arises, doubtless, from the featureless character of the fair goddess' face, indicative of the sullen thickness of the veil in which she is enwrapped,—we shall have an additional reason for concluding that life is also absent from this the second planet of the system.

[1] *Other Worlds*, pp. 74-78.   [2] *Ibid.* p. 78.

FONTENELLE, in his fourth evening, remarking that Mars afforded nothing curious that he knew of, hastily dismisses the planet by observing, 'Let us leave Mars, he is not worthy our stay.' But it has been upon Mars the hopes of modern pluralists have been most largely built. Indeed, Mr. Powell, with a curtness even greater than that of Fontenelle, and a confidence equal to that of Phipson,[1] dismisses it by saying: 'The planet Mars is on all hands admitted to be circumstanced so similarly to our Earth, that little discussion can be needed.'[2] More recent opinion, however, is not quite so unanimous. Mr. Proctor has, indeed, advocated its habitability at some length in a chapter of *Other Worlds than Ours*, and in another chapter of *The Orbs around us*, setting out by laying it down to be 'singular that among all the orbs which circle around the Sun,[3] one only, and that almost the least of the primary planets, should exhibit clearly and unmistakeably the signs which mark a planet as the abode of life.'[4] On the other hand, Mr. Maunder, superintendent of the Physical department at the Royal Observatory, Greenwich, in several papers contributed by him to the *Sunday Magazine* for 1882, after a discussion of the subject, says, 'We can scarcely suppose that any living beings can exist upon a planet so unhappily circumstanced.'

[1] Phipson's *Mysteries*, p. 268.   [2] Powell's *Unity*, p. 232.
[3] *Other Worlds*, p. 84.
[4] The observation is practically repeated in *The Orbs*, p. 105.

While I shall be found to concur in a conclusion akin to that of the last named, I do not propose to adopt all the inferences of either writer, but extracting from both, just as from other sources, whatever facts can be elicited to bear upon the subject, I shall endeavour briefly to state the views which occur to me leading up to that conclusion.

And, first of all, it is proper to give those physical facts about which astronomers, differing, no doubt, slightly, as usual, in their figures, are agreed.

Perhaps more differences occur in stating the size of Mars than in other particulars. Mr. Proctor states its diameter to be 5000 miles; Lardner, only 4363 miles, and others a good deal less. It may be considered as a little over half the diameter of the Earth. But the diameter at the poles is less than the equatorial diameter by 1/16th, or about 250 miles,[1] while the difference between the two diameters of the Earth is only $26\frac{1}{2}$ miles, showing a much greater, I may say an immense flattening at the poles of Mars.[2]

Its density was supposed to be nearly equal to that of the Earth, and Mr. Phillips[3] considered it to be a hollow globe, 'with an average thickness of crust equalling 569 miles.' However, since the weight of the planet is now said to be 1/10th that of the Earth, the density, I presume, is to be considered very greatly less.

---

[1] Grant's *History of Physical Astronomy*, p. 236.
[2] The estimates made by different astronomers of the amount of flattening at the poles have varied much. See Guillemin's *Heavens*, p. 185.
[3] *Worlds beyond the Earth*, p. 122. He estimates the crust of the Earth may be 2095 miles thick, but not solid throughout.

It results that the force of gravity, which has an important bearing on the present question, is much less upon Mars than upon the Earth.[1] It is indeed less than half, one pound weighing only about six ounces, less or more, according to the authority stating it.

The weight has been ascertained by means of the two tiny satellites recently discovered. Their existence had been conjectured, but it was not till 1877 they were found. The outer moon, at a distance of 15,000 miles,[2] revolves round Mars in 30 hours 14 minutes. The inner one, at a distance of 6000 miles, in 7 hours 38 minutes, or $3\frac{1}{4}$ times in one day. The sizes of these small moons have not, I believe, been, and perhaps cannot be ascertained.[3] Mr. Proctor suggests that each of them may not be more than ten miles in diameter.[4] The two together cannot give 1/127th of the light of our Moon.

The inclination of the axis of the planet to the plane of its orbit differs a few degrees from that of the Earth, or, according to the authority taken, from 4 to 7 degrees more inclined from the perpendicular.[5] Its day in length resembles our own, being nearly 40 minutes longer than ours; but, consequently, as it is a smaller body, it revolves more slowly—nearly twice as slowly.

Its year is 687 of our days, or within six weeks of two of our years.

Its orbit, it is important to observe, is eccentric.

---

[1] Leitch, p. 49; *Other Worlds*, p. 87; Maunder, p. 104.
[2] *Knowledge*, i. p. 167.   [3] *Year Book of Facts for* 1878, p. 161.
[4] *Flowers of the Sky*, p. 187.
[5] Proctor, $3\frac{3}{4}°$; Herschel, 7°; Dick, 7°.

The mean distance of Mars from the Sun is about 145,000,000 miles, above 50 millions of miles more distant than the Earth; but when in aphelion it is $152\frac{1}{4}$ millions of miles, and when in perihelion it is $126\frac{1}{4}$ millions of miles distant from the Sun, so that there is a difference of no less than 26 millions of miles between its position when nearest and its position when farthest from the orb from which, in common with all the planets, it derives its light and heat. And it may be here noted that the effect of a far less eccentricity in the Earth's orbit continued during a long period is stated to have produced a glacial period. See Cooley's *Physical Geography*, p. 411.

When at mean distance from the Sun the light and heat he receives are less than we receive in the proportion of about 4 to 9, or, as Leitch expresses it in decimals, Earth being reckoned $= 1$, Mars $= 0.43$. Mr. Proctor tells us—[1]

'The axis of Mars is so situated that the summer of his northern hemisphere occurs when he is at his greatest distance from the Sun. The same relation holds in the case of the Earth, the Sun being 1,500,000 miles nearer to us in winter than in summer; whereas to those who live in the southern hemisphere he approaches nearer in summer than in winter. But the effects resulting from the relation in the case of Mars must be very much more striking than those we recognise. For whereas the Sun gives only one fifteenth more heat to the whole Earth in January than he does in July, the Sun of Mars gives half as much light again in perihelion as in aphelion. The summer of the northern hemisphere of Mars must be rendered much cooler, and the winter much warmer, by this arrangement. On the other hand, the contrast

[1] Proctor's *Other Worlds*, p. 88.

between the summer and winter of the southern hemisphere is rendered more striking than it would otherwise be.'

In consequence of the greater length of year, the seasons of Mars are nearly double in length those of our planet, the spring and summer of the northern hemisphere lasting 76 days longer than they do in the southern hemisphere, while by reason of the greater inclination of its axis the tropical and arctic zones extend farther, reducing correspondingly whatever of temperate zones there may be to a minimum.

Mars becomes favourably situated for observation only 'once in every two and a quarter years, continuing to be well placed for only a few months,'[1] and though observed for more than two hundred years, Mr. Proctor states that, 'taking into account all the requirements for good definition, it may be said that Mars has not been under really effective observation for more than a very few days.'

But the observation which has been made has discovered certain markings upon its surface which are not temporary, but form permanent features, so much so that a chart of the appearance of Mars has been made exhibiting green portions, red portions, and white portions. These have been described and considered to be seas, land, and ice-poles respectively, and I shall assume that they are correctly so designated, and take the map given by Mr. Proctor as truly representing what is so seen.[2]

---

[1] *Other Worlds*, p. 102.
[2] The history of the charting of Mars, and a description of the charter, will be found stated in Proctor's *Other Worlds*, p. 91, and more fully in his *Orbs around us*, pp. 107-115.

A theory was indeed propounded, at a meeting of the Royal Astronomical Society in 1877, by Mr. Brett,[1] that the glowing red colour of Mars is red-heat, so that the planet is not, he considers, in a condition similar to the Earth; while the white polar patch is not only not snow, but it constitutes no part of the solid mass of the planet, and is nothing more than a patch of cloud, the only real cloud existing in Mars. But this view does not appear to be approved by other observers, and as there seems really nothing either to confirm or to warrant it, it may, I think, be thrown out of consideration; although, were it established as a fact, it would effectually exclude the thought of Mars being habitable.

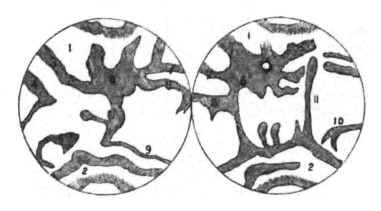

CHART OF MARS.

1. *Lockyer Land.*
2. *La Place Land.*
3. *Hooke's Sea.*
4. *Maraldi Sea.*
5. *Dawes' Ocean.*
6. *Delarue Sea.*

7. *Arago Strait.*
8. *J. Herschel Strait.*
9. *Nasmyth Inlet.*
10. *Huggins' Inlet.*
11. *Besset Inlet.*

The above, then, is a rough reduction from Mr. Proctor's chart of Mars, given in *Other Worlds than Ours*, which was so very beautifully executed as really

[1] *Year Book of Facts for* 1878, p. 161.

to suggest it to be 'too good to be true.' The shaded portions, intended to represent what are said to be ocean and sea, are rather heavier or larger in this small scale than in the original. How very different the representations of Mars are in other books may be observed upon comparison, as, for example, with those presented in Guillemin's *Heavens*, p. 181.

Taking, therefore, Mr. Proctor's chart, here introduced on a smaller scale, for convenience of reference, as exhibiting what have been considered as land, seas, and ice-poles, the first observation which it elicits is the vast difference between the superficial extent of sea and land upon Earth and upon Mars. On Earth the sea covers nearly three-fourths of its surface. In Mars the proportion of land is much greater. Williams[1] estimates the proportion of land to water as five times greater than the proportion of land to water on the Earth. But calculations widely differ, although by the smallest computation land and sea are equal. Mr. Lockyer, in one of his books, computes it at four times the extent of sea. Mr. Proctor says: 'There is little disparity between the extent of oceans and continents,'[2] and in another book,[3] mentions he 'cut out with a pair of scissors the parts representing land and the parts representing water (leaving the polar parts as doubtful), and carefully weighed them in a delicate balance.' He 'found that they were almost exactly equal; whatever preponderance there was seemed to be in favour of the land.' Of course the accuracy of this computation depended entirely on the accuracy of the map so weighed.

[1] Williams' *Fuel of Sun*, § 250.
[2] *Orbs around us*, p. 116.
[3] *Flowers of the Sky, the Planet Mars*, p. 183.

Which of these two authorities is correct? Perhaps the truth lies between the two, in saying that in superficial measure the land of Mars is double the extent of sea.

But the mode of distribution is very different. The sea is not, as on Earth, one body, but seems divided into separated portions, and, in particular, the arctic seas are separated from those adjoining by belts of land: La Place Land at the North, and Lockyer Land at the South Pole.

Further, the seas are not only peculiarly shaped, but they are for the most part laid out in long narrow strips or inlets, such as Nasmyth's Inlet, Huggins' Inlet, Besset's Inlet, Arago and Herschel's Straits, and the Hooke and Miraldi Seas.

Not only so, but when it is considered how great is the flattening of the poles of Mars,—for according to Herschel it is 1/16th, and to Arago, 1/32nd of its diameter (though the latter[1] believes it to be more), or causing a depression of say from 60 to 120 miles at each pole,—we might expect, upon the principle that the seas ought to flow to the lower level, not only that they should have been connected between the polar and the other seas, but that the whole water of the planet would have been concentrated at the poles. It is not so, and hence it probably is that what are called seas, if seas at all, are merely shallow basins or lagoons; and it lends countenance to the supposition to bear in mind that the surface, covered in winter with what appears to be snow or ice, rapidly loses this covering with the advent of summer, indicating a very thin layer. It is in striking contrast with our poles, and especially our South Pole,

[1] Humboldt's *Cosmos*, p. 503.

which in the height of summer are extensive unapproachable ice-fields.

If the seas be shallow, the land is not likely to be mountainous. Probably there are only elevations sufficient to hem in the seas; and this seems the more probable because the land does not exhibit shining peaks, which, if they existed, would be covered, or sometimes covered, with snow or ice. The only exceptions appear to be two,—a spot in the middle of Delarue Sea, and another spot in Dawes Ocean,—which in all likelihood are mountains, and that not of any height, as, if the brightness occasionally seen be attributable to snow, it is not permanent.[1]

As the seas occupy so much smaller surface than on

---

[1] I have stated this on the authority of Mr. Maunder,[2] who says: 'On September 29 (1877), Mr. Green, an astronomer, who was watching the planet at Madeira, noticed that a large part of it was no longer dark, but bright. Three days later an Italian astronomer saw that the greater part of the spot was dark as usual, but that a *white spike* jutted out into it from the neighbouring lighter markings. *Two or three days later these had vanished*, and the dark spot had resumed its ordinary appearance. Another large spot, called after a clergyman named Dawes, noted for his keen sight and the skill with which he effected the most delicate observations, is very frequently concealed from view in the same manner. Thus in 1877 it was generally seen in its entirety, dark and distinct; but in 1879, as a rule, only portions of it could be caught here and there, the greater part appearing as bright as any part of the general disc.' If these be mountains with snow upon them, their elevation must be low, and the snow remarkably thin and evanescent. Mr. Proctor does not entirely differ, but states: 'In Delarue Ocean there is a large island which presents so strikingly brilliant an aspect that it has been supposed to be covered (ordinarily) with snow. It has been called Dawes Ice Island,' Proctor's *Other Worlds*, p. 93. Mr. Proctor does not notice the other island or mountain.

[2] Maunder, *Sunday Magazine* for 1882, p. 32.

the Earth, the land stretches in extensive and oddly-shaped continents, and has a ruddy appearance, so much so that Mars shines a ruddy star. With this ruddy appearance I shall have immediately to deal.

The bright spots at the poles of Mars were carefully observed by Sir William Herschel,[1] who supposed them to be occasioned by light reflected from masses of ice and snow accumulated in the polar regions,—an opinion which, though it has been doubted, is now generally held. The remarkable circumstance connected with them is, that they increase in extent so largely during the winter months, and diminish so rapidly during the summer months. The difference between the two extremes is as much as 35°. Mr. Maunder[2] says the Cap has been watched until it was reduced to only 130 miles in radius; and in 1877 the South Pole was actually free from ice. Sometimes the one pole,[3] however, and sometimes the other is the brighter, or is the more extensively covered. But if our Earth were covered with snow and ice as extensively as is Mars during winter, the covering would extend over nearly all Europe, embracing the south of France and the whole northern half of Asia; while in the Southern hemisphere it would cover New Zealand and part of South Australia, reach into South America to Buenos Ayres, and in Africa to the Cape of Good Hope.

And now the atmosphere of Mars claims consideration. For that it has an atmosphere of some kind seems to be admitted.

Dr. Dick[4] states: 'From the gradual diminution of

---

[1] Grant, pp. 236, 237.
[2] Maunder, p. 103.
[3] *Ibid.* p. 31.
[4] *Celestial Scenery*, p. 119.

the light of the fixed stars when they approach near the disc of Mars, it has been inferred that this planet is surrounded with an atmosphere of great extent;' although he mentions that the extent has been much overrated.

Mr. Proctor [1] tells us, that 'towards the edge of the disc the ruddy and the greenish tracts are lost in a misty whiteness, which grows gradually brighter up to the very border of the planet;' that 'a veil is sometimes drawn over it for hours or even days together;'[2] and that 'it has been noticed that when it is winter in one hemisphere, and therefore summer in the other, the former hemisphere is nearly always hidden from view by just such a veil as I have spoken of above;'[3] and this veil he considers to be clouds formed just as our clouds are.[4] Mr. Maunder, on the other hand, says: 'The clouds generally seen are only sufficient to lighten a little, not to fully hide, the dark markings, and do not reflect nearly so much light as our own heaven-clouds do.' And this is what we may readily suppose from the nature of things at Mars. There are clouds, therefore, but not to the same extent or having the same density as on Earth, and clouds indicate both moisture and air.

We should, however, be left in some uncertainty with regard to this atmosphere were it not that the spectroscope has come to our aid.

Mr. Roscoe,[5] referring to the observations of Dr. Huggins, Dr. Miller, Padre Secchi, and Janssen, says:—

---

[1] *Other Worlds*, p. 90.   [2] *Ibid.* p. 96.   [3] *Ibid.* p. 97.
[4] See Mr. Proctor's views as to clouds on Mars, *Orbs around us*, pp. 120, 121.
Roscoe's *Lectures*, p. 231.

'From observations upon Saturn, it appears probable that aqueous vapour exists in the atmosphere of this planet as well as in that of Jupiter. In Venus no intensifying of the atmosphere could be observed; but some remarkable groups of lines, corresponding to those seen when the Sun is low, were noticed on the more refrangible side of the line "D" on the Mars spectrum; and these indicate the existence of matter similar to that occurring in our own atmosphere.'

Secchi and Janssen conclude that in all probability the vapour of water exists in the planetary atmospheres.

There is appended to Mr. Roscoe's lectures[1] an article by Dr. Huggins, 'On the spectrum of Mars, with some remarks on the colour of the planet,' in which he observes to the same effect; and after showing that the ruddy tint of Mars is not derived from its atmosphere, he[2] gives us these important facts: that Mars resembles the Moon in the variation of light reflected from it; that its mean reflective power '*is not more than one-half greater than that of the lunar surface;*' '*that the light is reflected almost entirely from the true surface of the planet,*' in contrast with Jupiter and Saturn, where the light 'has evidently come from an envelope of clouds,' and which 'have an Albedo severally about four and three times greater than that of the Moon.' The Albedo, as appearing from the observations of Zöllner, being as follows:—

Moon	=	·1736	of the incident light.
Mars	=	·2672	,,
Jupiter	=	·6238	
Saturn	=	·4981	
White Paper	=	·700	
White Sandstone	=	·237	

So that Mars reflects just a little more strongly than

[1] Roscoe's *Lectures*, p. 266.   [2] *Ibid.* p. 268.

white sandstone, probably indicating a character of surface akin to it.

Dr. Huggins thinks it highly probable that the conditions of surface which give rise to the phenomena to which he refers are common to the Moon and Mars.

But the importance of the observation upon this reflection of light from Mars is, that it indicates a very slender or rare atmosphere surrounding the planet; for we have seen that the Moon is altogether devoid of atmosphere, and the condition of Mars must in this respect approach that of the Moon.

The atmosphere[1] itself has been said most probably to resemble that of the Earth in general constitution, and it may possess the elements which enter into the composition of our atmosphere, but we cannot, as yet at least, tell whether they are combined in the same proportions, or whether it may not contain, as very likely it does, other elements fatal to life. The mere proof of the existence of an atmosphere does not necessarily establish it to be fit for the support of life. A slight excess of one element alone may be sufficiently adverse.

Dr. George Wilson, writing in 1855, referring to Mars, says:—

'A globe with water and an oxygen atmosphere might certainly put in some chemical claim to be a sister of the Earth. But such speculation is premature. The presence of water does not justify the inference that free oxygen is also existent; nor does it warrant the conclusion that more than fifty other elements must be there also.'[2]

All the circumstances I have now stated combine with each other in militating against the habitability of

---

[1] Proctor's *Orbs*, p. 122. [2] *Chemistry of the Stars*, p. 24.

Mars. The smaller size and smaller gravitating power; the greater distance from the Sun, and unusual eccentricity of orbit; the greater extent of land surface, and lesser extent of sea surface, and their configuration; the rarity of the atmosphere,—all have an effect, and inseparably act and re-act on each other.

The greater distance of Mars from the Sun, combined with the smaller extent of sea surface, ought to have the effect of producing less amount of evaporation from the seas, and of diminishing the size and density of the clouds of Mars, which will rise slowly and reach greatly higher; and though expanding more as they rise, and in expanding losing heat or cooling more than the clouds of Earth. They will therefore afford a lesser covering to the body of the planet. The result of this must be that the heat which it receives during day will more freely irradiate into space through the night than happens upon Earth, so that any heat which it may gain from the Sun by day will be largely lost by night; and during night at least the cold which results from the greater distance of the planet from the Sun will necessarily be greatly increased by this cause.

Mr. Williams says[1] that as the intensity of the Sun's heat upon Mars is to that of the Earth as 0·431 to 1, the mean temperature

'must be considerably below the freezing-point; from the thinness of the atmosphere, and the small quantity of aqueous vapour it must contain, the variations of temperature between day and night must be very great.' 'The seas must be frozen to the bottom; but the surface of the water on all parts of the planet, which are exposed

[1] *Fuel of the Sun*, p. 167.

HEAT. 263

with only moderate obliquity to the Sun's rays, would be thawed to a depth varying with the duration and verticality of such exposure. *Its surface would thus be thawed during the day and frozen again at night*, like the surface of our own Alpine glaciers, but the variations would be far more sudden and severe on the Martial ocean.'[1]

And Mr. Williams thinks there should be a feathery dew of hoar frost beginning to fall before sunset, and a current flowing towards the dark side of the planet corresponding with this, the outer portions being more luminous.[2]

Mr. Maunder, in discussing the habitability of Mars, points to the anomaly of a long narrow sea, called Nasmyth's Inlet, situated N. lat. 43°, not being apparently frozen in winter like the Caspian and other seas, in corresponding latitudes in the Earth, as a circumstance which would lead to the inference that Mars has as warm a climate as the Earth, and that something on Mars tends to counterbalance the effect of distance of the Sun. This cannot, he shows, be either the result of internal heat or of greater density of the atmosphere. To maintain as much heat on Mars by means of atmosphere as is enjoyed by the Earth, Mars would require to have six times as much air as we have above every square inch, and 'this vast atmosphere would extend not merely six times as far from the planet as our own, but nearly thrice as far again as that.'[3] This (which would be equivalent to 900 miles deep) could not but be perceivable,[4] and not only would obscure or hide the markings of the planet, but would

[1] Williams' *Fuel of the Sun*, p. 168. [2] *Ibid.* p. 178.
[3] Maunder, p. 104. [4] *Ibid.* p. 170.

reflect far more light than it does. On the contrary, it is the case that the atmosphere is meagre and as rare as it is on the tops of our highest mountains, 'and yet these peaks are lifeless solitudes even when placed in the tropics, the hottest regions of the Earth; how much more, then, would it be impossible for anything to live when to so rare an air is added the far more intense cold that must prevail in the lands of distant Mars.'[1] Mr. Maunder then explains that the Sun's rays would pass through the rare atmosphere without warming it, but would warm the land, and cause evaporation on the sea, and the vapour exhaled would be better kept, and it would hang on the sea whence it is derived, and prevent the radiation of heat, and might so keep the seas from freezing. But Mars when observed by us is seen at mid-day, while under the power of the Sun's rays, while in winter it is badly placed for being seen by us. The conclusion, then, apparently to which Mr. Maunder comes (if I have rightly apprehended his meaning) is, that from its position when we see Mars we see it under sun heat, and it appears warmer than it really is. Seas which seem to us to be unfrozen may be frozen when not under our vision, or during night and during winter. His words are: 'So the parts of Mars which we see best at any time are just those which are for the interval the warmest; as if the stout-hearted little planet, true to his warrior reputation, strove always to hide his calamities, and put the best face upon his rather unpropitious circumstances.'[2]

Mr. Maunder's view seems corroborated by a fact stated by Captain Scoresby (*Journal*, p. 291), that in clear, calm weather the sea in the interstices of the ice generally freezes on the *decline of the Sun towards the*

[1] Maunder, p. 170.   [2] *Ibid.* p. 172.

meridian below the pole, though the temperature be 32° or higher.[1]

But may we not rather adopt the supposition, consistent with the fact that water being a non-conductor of heat, so that the lower portions of the seas must always be cold,[2] that the seas are constantly frozen, and 'consist of pure ice throughout, retaining the partial transparency and greenish colour of deep-sea water.'[3]

Whichever view be taken seems to justify the idea that the seas of Mars, if they resemble our own, are very shallow.

We now come to the important bearing of the excessive cold in Mars on life. Skertchly[4] says:—

'If our Earth were removed to the orbit of Mars, all life would be destroyed in consequence of the intense cold that would prevail. The mean distance of Mars is about twice that of the Earth, consequently an equal area on the surface of Mars will only obtain one-fourth of the amount of heat received from the Sun upon such an area on Earth. We may take, for instance, the mean annual temperature of England at 50° F.; then at the distance of Mars its temperature would be 12·5° F. or 19·5° F. below freezing-point, which is considerably below the mean temperature of any known place, and about equal to the temperature of South Greenland in January. Similarly, the mean temperature of the equator is about 80° F.; at the distance of Mars it would be 20° F. or 12° F. below freezing-point, which is about the mean temperature of Novaya Zemlia. Even the summer temperature of the Sahara, which is now 95° F., would fall to 23·7° F., or 8·3° below the freezing-point.

---

[1] Daniel's *Meteorology*, ii. p. 132.   [2] Herschel's *Meteorology*, p. 38.
[3] *Flowers of Sky*, p. 181.   [4] Skertchly's *Physical System*, p. 136.

'But these calculations do not show the full extent of the cold we should suffer at the distance of Mars. As the oceans would be frozen, no currents could convey heat from the tropics to the polar regions, which would, in consequence, have their climates vastly more rigorous than has been assumed. If we are right, then, in calling the white polar patches on Mars ice-caps, it follows as a necessary consequence that the physical conditions of that planet are very different from those which prevail on Earth.'

Mr. Maunder rests his conclusion mainly on the fact of this extreme cold.

'But winter!' he says, 'when even the Sun's noonday heat might not suffice to raise the thinnest vapour, and when, as soon as the short day was over, the scanty heat received during its few hours would be radiated off into space through the dry and chill air without let or hindrance, How will it be with the red planet then? Cold, far colder than our poles, far colder than our loftiest mountain tops, colder than aught of cold of which we know, cold such as nothing living might endure, must reign over the sunless hours of the Martial winter. Nor would that winter last only for a few short weeks, but more than three hundred days would roll by, and still the ice king would sway an unchallenged sceptre over the unhappy regions under his control.

'But when the summer broke out at last on those frozen realms, what mighty changes would take place! The winter's snow would melt, the imprisoned rivers be released, and floods caused far more extensive indeed than any we ever experience here—so extensive, indeed, that their traces can be clearly perceived across fifty millions of miles. For a period considerably longer than our entire year, the Sun beats on the northern pole of Mars without once sinking below the horizon; and when we remember this we cannot be surprised that his polar snows should disappear with the rapidity which we actually notice, nor that such immense quantities of water vapour should be raised into the air as to make known its presence to us by means of the spectroscope at this immense distance.

'We can scarcely suppose that any living beings can exist upon a planet so unhappily circumstanced. To endure such long and bitter cold, to breathe in so rare an atmosphere, to escape deluges so mighty, would be impossible to forms of life like those we see around us. Movement and change there may be in abundance, but it will be the change of inorganic matter. Winds will blow, clouds gather and disperse; it may be that earthquakes will rend the rocks, or volcanoes scatter an ashen rain over miles of country, but the wondrous changes of life in all its infinite variety of form will never be noticed there.'[1]

Such are Mr. Maunder's conclusions. But I think we can go a good deal further.

The long narrow sea to which he refers evidently lies in a fissure of the planet, and it may be that this is a deep and dry fissure where there is no water to freeze, and hence it would not shine as if it had been frozen.

But this is a supposition far from probable, and there is another more likely. We know that salt water does not freeze so readily as fresh. May it not, therefore, be that this inlet (together with others similarly situated) contains a sea resembling the Dead Sea in Palestine, or even composed of or containing ingredients still less liable to freeze,—a sea which may not resemble any sea upon Earth, and one, it may be, containing water in it, but water mixed with the most deadly ingredients.[2] It will be kept in view that the proba-

[1] Maunder, *Sunday Magazine*, 1882, p. 172.

[2] Mr. Proctor argues that because of the detection of aqueous vapour by the spectroscope, the seas must consist of water, and not of other

bilities of such a case are great, for—(1) neither Sun nor Moon has power sufficient to create tides in Mars;[1] (2) the seas not being united as on Earth, there can be no currents flowing through them such as flow through our seas; (3) if one authority be correct, the winds on Mars can never be sufficiently violent to produce storms, at all events, they are more gentle than with us;[2] (4) that there can be no heavy rains; (5) there are probably no rivers. Lastly, the sunlight and heat, owing to greater distance, being less than half what are received by the Earth, must create a difference of influence on the water which, under the preceding circumstance, can only be stagnant. Perhaps the progress of science may reach the point of discovering of what ingredients these so-called seas are composed; but it is quite open for us at present to suppose, and not unreasonably, that the seas are unfavourable to life. We know the following to be scientific facts: (1) that in seas and lakes which have

---

fluids (*Other Worlds*, pp. 102-104); but this does not by any means seem to follow. Vapours rise from the Dead Sea, and densely too, but the salt remains—only the water rises.

[1] 'The effects of the Sun in producing tides must always be inappreciable in Mars' (*Orbs around us*, p. 117).

[2] One cannot, indeed, speculate much regarding the winds in Mars beyond this, that there being less moisture, there will be less room for winds. The theories regarding the origin of winds differ. See Hopkins' *Atmospheric Changes*, pp. 40 et seq. and 313 et seq. But Mr. Proctor states that 'no winds would disturb the surface of the frozen seas, for winds have their origin in heat, and with the death of the solar heat the winds would utterly die out also' (*Flowers of Sky*, p. 181). Guillemin, on the contrary, in stating that notable differences distinguish the Earth and Mars in their meteorology, refers to a remark of Professor Phillips, that 'the considerable periodical exchange of moisture which is made between the two hemispheres (of Mars), especially between the two poles, must give rise to hurricanes and storms of the violence of which we can form no idea' (Guillemin' *Heavens*, p. 184).

no outlet, and where, consequently, all the water passes off by evaporation, the greatest degree of saltness is attained, Buchan's *Meteorology*, and examples there cited, § 180; (2) that the density of seas is dependent on the quantity of fresh water poured into them from rivers, Buchan, § 181; (3) that the density is increased or diminished according to the lightness or heaviness of the rainfall, Buchan, § 182. But there is another little circumstance which may not be without its value. Mr. Proctor describes the seas as 'being for the most part darker' (than the land) 'and of a somewhat *greenish hue.*' Now the Dead Sea, as described by Robinson in his *Biblical Researches*, when looking down upon it from a lofty height, 'appeared decidedly green, as if stagnant, though we afterwards saw nothing of this appearance from below.' This may help, together with the other circumstances, to suggest a resemblance to the Dead Sea, of which it is said, according to the testimony of all antiquity and of most modern travellers, 'no living thing—no trace in fact of animal or vegetable life—exists in this sea.' And this, chemical analysis has shown, results from the matters contained in it differing greatly from those of sea water.[1] The great Salt Lake of America is similarly fatal to animal and vegetable life, and we are told that not a tree grows on the shores of the lake nor on the adjacent plains.[2]

Passing from the supposed seas to the land, the first observation which occurs is, that it presents a ruddy appearance or colour. This has been established by Dr. Huggins not to be due to an absorptive power in

---

[1] Elisée Reclus' *Earth*, p. 476.
[2] *Ibid.* p. 478. See also Mrs. Somerville's *Physical Geography*, ii. p. 21.

the atmosphere of Mars, and Sir John Herschel states that[1] 'it indicates, no doubt, an ochrey tinge in the general soil, like what the red sandstone districts on the Earth may possibly offer to the inhabitants of Mars, only more decided.' Mr. Maunder, telling how the Earth would look as seen from Venus, says: 'The great sandy deserts, Sahara and Arabia, would appear to be fully as ruddy as anything we see on Mars. It is hard to say what colour England would show, with its green woods, and its golden cornfields, and its black mining districts; probably a pale greenish yellow.'[2] Now there is not a particle of green visible on these lands of Mars, nor is there, as already stated, any variation from time to time upon its hue, so that the conclusion is most natural that there is no vegetation of any kind on its surface. The land may be iron oxydised, but more likely is either rock or desert sand, and there are various circumstances which go to such an inference.[3]

(1.) That the Albedo or reflective power of Mars so closely approaches that of the Moon, the Albedo of the Moon being ·1736 and that of Mars ·2672.

(2.) The great extent of land surface compared with sea surface—looking to the very different relations they bear to each other on Earth—shows that, even were the evaporation of the sea equal to the evaporation of the waters of Earth, the moisture which the land would receive would be many times less than the moisture received by the land upon Earth,—probably it would

---

[1] § 510.   [2] Maunder, p. 32.

[3] Dr. George Wilson says: 'The strange fiery red light of this star also implies a peculiar condition of its whole uncovered surface, very unlike what our Earth's exterior exhibits, and forbids any conclusion as to the general identity of their superficial condition ingredients. It still more forbids rash inferences as to terrestrial plants and animals existing on a body of unknown composition,' p. 24.

not be one-tenth of what the Earth receives, and therefore altogether insufficient for the support of vegetation.

(3.) But, as already said, this moisture would not fall in showers of rain, it would be merely at best a moist mist, which would not reach the interiors of the continents. If there be no mountains there would be no rivers, nor would rivers be possible under the circumstances just mentioned. It may be said, therefore, there would be no fresh water to support life or vegetation.

(4.) The excessive cold of winter would destroy all vegetation, if any did exist. 'It is pretty generally believed that if the temperature fall below 40° the growth of vegetation is arrested during the next day.'[1] But as the solar heat upon Mars is of itself less than half that of Earth, it is apparent that there could be no vegetable growth on Mars. The effect of the severe cold on Mars, however, may be still greater. It is related that, in an attempt to sink a well in Siberia,[2] it was found that the Earth was frozen to a depth of 612 feet, although regarded as an aqueous rock, protected from change beyond the depth of a few feet by the shortness of the summer and the general rigour of the climate.

(5.) But vegetation would also be destroyed by the extreme and sudden transition of warm to cold, and cold to warm, during the days and nights. Mr. Buchan says, § 113 :—

'In the north-western parts of the United States of America the temperature in spring often rises to 83° during the day, and falls to freezing during the night. Under such a climate the vital functions of plants, the tissues of which abound in sap, are called into activity

---

[1] Buchan, § 119.   [2] Cooley's *Physical Geography*, p. 82.

during the day, but the sap being frozen during the night the vessels containing it are ruptured by expansion, and the plant, if not totally destroyed, is so seriously injured that its successful cultivation becomes precarious and uncertain.'

Now on Mars, where the air is so rare and evaporation is so small, it is obvious that the change of temperature from day to night and night to day must far transcend anything we are acquainted with, and hence the greater impossibility of vegetation existing on Mars.[1]

(6.) The absence of vegetation, and especially of forests, would of itself have the effect of depriving Mars of the modifying influences they bring to climate; vegetation producing a cover which modifies the heat of day and the cold of night, and equalizes the seasons.[2]

(7.) But if Mars cannot support vegetation, then the food of animal life is wanting.

Lastly, the very same causes which would be detrimental to vegetable life would also be fatal to animal life. For the severity of the strain on the constitution from an excessively warm climate to an excessively cold one, from day to night and night to day, and from winter to summer and summer to winter, and the more so that there would be little twilight and no spring or autumn, and only a narrow temperate belt, would be such of itself as would be fatal to life. Add to this the want of fresh water, and the rarity of the air, and the absence of vegetable substance, and we shall see the impossibility of animal, much less of rational animal life upon Mars.

[1] Mr. Proctor has pointed out the effects of a greater length of year and of the lesser gravity in Mars on vegetation in his *Expanse of Heaven*, p. 67 et seq., to which reference may be made.

[2] Buchan, § 132.

VII.

BEFORE proceeding to the consideration of the Major Planets, there are some bodies appertaining to the solar system, and having a link of connection with each other, which have claims to be treated by themselves.

1. THE ASTEROIDS AND METEORIC STONES.

The Asteroids, or as they are sometimes termed, the Planetoids, or the Telescopic Planets, number, so far as already detected, 230 or more, and additional ones are being yearly discovered, and they are believed to extend to many thousands in number. According to the calculations of M. Leverrier, they may even amount to 150,000.[1] The history of the discovery of the earlier ones, commencing with Ceres, on the 1st day of the present century, will be found in Grant (p. 238 et seq.). Among the popular works, particulars are to be more fully had in Dr. Dick's *Celestial Scenery* than elsewhere, although necessarily the information at his command was limited to the date at which he wrote.

Great diversity of calculation has been made regarding their sizes. Schroeter estimates at figures much higher than those of Sir William Herschel, or than those of recent astronomers. Thus Schroeter considered Ceres to have a diameter of 1624 miles, and, including a supposed atmosphere, 2974 miles. Pallas to have a diameter of 2099 miles. But probably the largest of

[1] Guillemin's *Heavens*, p. 192.

these small planets is not 700 miles in diameter; while Herschel's view was that none of them exceeded 163 miles;[1] and Humboldt's, that probably 'the largest of the small planets is at the utmost 145 geographical miles in diameter.'[2] But many of the smaller planetoids must be exceedingly minute; nay, the possibility is that there may be multitudes, perhaps millions, which are far too microscopical for discovery by the telescope. Indeed, even the larger ones, except occasionally Vesta, can only be discerned by the aid of a telescope of high magnifying power.

Now those who are advocates of plurality must surely admit a limit in point of size to the possibility of a planetary body being inhabited. A body having the diameter of a yard or a mile, whatever might be its condition or position, could not be supposed to be the scene of rational life; and on this ground exception would fall to be extended to the most of these planetoids.

But the power of gravity being proportionally so weak, would of itself be a determining force. On many of these bodies, perhaps on the vast majority of them, no small animal, no small object, would be safe. There must be wanting a power of gravitation sufficiently attractive to retain them on its surface. Exposed, as most of them are, to a movement in their orbit at the rate of upwards of 40,000 miles per hour, probably anything slight, or even anything unattached, might be blown or expelled from their surface, or in any view, might be easily drawn from it by counter attraction, such as by the attraction caused by one of themselves: for in a manner strange and unique, and doubtless, were they inhabited, terrific, they cross each other's orbits, and

[1] Dick, *Celestial Scenery*, p. 132.   [2] *Cosmos*, iv. p. 424.

thus occasionally pass in dangerous proximity to each other.

We know not the shapes of these bodies, or whether even the larger ones are globular, or whether they revolve on axes, or if they revolve on axes, whether these are inclined to the ecliptic, and if so, at what degrees. If the facts were known, they might have a bearing on the question.

But this we do know, that their distance from the Sun is about, on an average, 100 millions of miles greater than that of Mars, and the light and heat they receive may be on an average a sixth or a seventh of that which is enjoyed by us upon Earth, so that their climate must be cold, vastly colder than what we have seen to be the unendurable coldness of Mars. But over and above, they are remarkable for the great eccentricity of their orbits. Mars, as already seen, has a difference of 26 millions of miles between its perihelion and its aphelion; but while a few of the asteroids have a less eccentricity than that of Mars, nearly all the others have a vastly greater. Thus Pallas is nearer the Sun at the one extreme than at the other by 129 millions of miles. The difference in the case of Euridice (discovered in 1862, after the date of Dr. Dick's book) is 20 millions of miles more. The distinctive extremes, therefore, of heat and cold in the course of their orbits must be beyond our conception; and all that has been said with reference to Mars on this subject applies with greater force to the asteroids.

There is something noticeable too with regard to the appearance of some of these small planets. Thus Vesta sheds a light more intense and white than do others, some of which are surrounded by nebulosity, and yet it is devoid of nebulosity. What can this mean? Can it be that it reflects from a metallic surface, and is therefore incapable of supporting vegetable and consequently animal life?

Then some of them which are said to have a nebulous surrounding possess it of great density, insomuch that Schroeter (who, however, seems to be here unreliable) estimates the atmosphere of Ceres to be 675 English miles in height, Pallas about a third less. What are these nebulous atmospheres? Are they remains of the original nebula out of which they were formed? or are they portions detached from passing comets? Or are the planets possessing such atmospheres, small as they are, not even yet cooled down to solidity? It may be so. Sir William Herschel[1] says that in viewing Pallas he could not

'perceive any sharp termination which might denote a disc, it is rather what I would call a nucleus. The appearance of Pallas is cometary, the disc, if it has any, being ill-defined. When I see it to the best advantage it appears like a much compressed, extremely small, but ill-defined planetary nebula.'

Sir John Herschel,[2] who conjectures that this appearance has possibly originated in some imperfection in the telescope employed, or some other temporary cause of illusion, thinks that the hazy appearance seen may only be 'indicative of an extensive and vapourous atmosphere, *little repressed and condensed by the inadequate gravity of so small a mass.*' But such a nebula as has been supposed to exist could not correspond with an atmosphere to support life. It is too dense, too opaque, too close to render life, or in any view, rational life,

---

[1] Dick, p. 131. [2] § 525.

possible. It would exclude the light of the Sun, feeble at the best. There could be no possibility of vegetation.

But beyond all that has now been stated, we are furnished with another fact which seems to bear, with overwhelming force, against any supposition for life in the asteroids, and it is derived from what we know of the meteoric stones which have fallen upon Earth. For meteoric stones may be regarded as only minute planets, such as many of the asteroids (which, from their minuteness, have not been discovered) must be,—planetoids which are not to be found in all regions of the solar system. 'These stones,' says Professor Leitch,[1] an advocate for plurality, ' are now almost universally held to be little planets; and they certainly afford a presumption that the other planets have similar chemical elements;' and in supporting his cause, he advances an argument upon them to which that of Dr. Wilson, to be immediately mentioned, is an unanswerable reply.

Accounts of the various falls of such stones will be found in Humboldt's *Cosmos*, vol. i. p. 103, and vol. iv. p. 587; also in Phipson's *Mysteries of Nature*, in his chapter on Aerolites.

Now the late Professor George Wilson,[2] who must be taken as an authority on the subject, tells us that

'the meteorites have been put upon the rack by the chemist, and all their secrets have been tortured out of them, but they have revealed fewer marvels than at one time was expected.' 'They contain no ultimate chemical compound which the Earth does not contain.'

[1] P. 300.   [2] *Chemistry of Stars*, p. 28.

And taking up the author of *Vestiges of the Creation*[1] where he states that

'they contain the ordinary materials of the earth, but associated in a manner altogether new, and unlike anything known in terrestrial mineralogy,' contriving, 'by an unwarrantable concealment, to convey a very false impression of the true nature of meteoric stones.' They are said to 'contain the ordinary materials of the Earth, which, no doubt, they do, but it should have been added that they contain only *some* of them, so far as we know, but the smaller part.'

And Dr. Wilson adds that

'so far as our knowledge extends, it does not appear that a third of our earthly elements has been found in these bodies. Humboldt, in his *Cosmos*, quoting from Rammelsberg, the greatest living authority on the subject, enumerates only 18 of the 60 elements as occurring in them. Professor Shepard counts 19 as certain, and adds two more as doubtful. It is to be observed, on the other hand, that not only are the majority of the terrestrial elements, including many of the most important among them, totally wanting from meteoric stones, but those which are present are not mingled, as the quotation (from *Vestiges*), indeed, acknowledges, in earthly proportions. Our globe consists, speaking generally, of two opposite classes of ingredients, namely, metals and non-metallic bodies, some of which, as oxygen in the one division and the precious metals in the other, occur free, but the greater number in combination with some body or bodies of the unlike class. There are many more *kinds* of metals than of non-metallic substances, but the latter, taken as a whole, occur in much larger *quantities* than the former. One non-metallic body alone, oxygen, is computed to

[1] It would appear from Dr. Wilson's *Life* that he had delivered a course of lectures on the *Vestiges*, which unfortunately was never published.

form a third of the weight of the crust of the Earth. In meteoric stones, on the other hand, whilst non-metallic elements are the less numerous constituents (only a half of those occurring on the Earth being found on them), they also occur in much smaller quantities than the metals. Of some of them, indeed, traces only are found.

'Many of the best marked aerolites are masses of *nearly pure metal*, chiefly iron, with a small proportion of nickel. Others contain cobalt, manganese, chromium, copper, and other metals, diffused through them in minute quantities, associated with a small percentage of oxygen, sulphur, chlorine, etc. The stoney meteorites consist chiefly of silica and metallic oxides.

'Whilst thus meteoric stones contain only a portion of the elements of the Earth, that portion is made up (in the greater number of meteorites), so far as the relative quantities of its components are concerned, almost entirely of metals. A meteoric stone represents, therefore, only a third of the whole constituents of the Earth, so far as number is concerned, and except to a small extent, but one class of them, so far as nature. A globe so constituted *could never* by any process of development (unless its so-called elements suffered transmutation) *become possessed of water, or an atmosphere, or give birth to terrestrial plants or animals.*'[1]

Dr. Wilson[2] then illustrates, in a characteristic manner, the consequence to a human being of the absence of only one element.

The fact of the far greater density of the smaller planets as compared with the greater ones, and the still greater density of the meteoric stones, lends countenance to the view that the asteroids, which are all small, though differing in size, are of a density or hardness, in all likelihood metallic, which precludes the possibility of vegetation and life.

[1] *Chemistry of Stars*, p. 30. [2] P. 32.

I cannot, however, leave the meteoric stones without noticing what was thought and has been represented as a momentous recent discovery, evidencing life on other celestial bodies. For it went forth that organic remains had been found in meteoric stones, enabling us 'to see with our own eyes the veritable remains of animate beings from another celestial body.' Two Germans, Dr. Hahn, a geologist, and Dr. Weinland, a zoologist, have, it was stated,

'recently investigated the subject to some extent, and the result of their labour is that they find in these stones fossil sponges, corals, and crinoids. They are also of opinion that they have discovered a trace of vegetable remains. These startling discoveries point to the fact that in the world where these rocks and animal forms originated, the cause of organic evolution was only similar to that which has taken place upon Earth.' The stones themselves were stated as 'believed by astronomers to be the remains of a planet which had been destroyed in some manner, shattered into fragments by forces which to us may be set down as unknown or mysterious.'

Elsewhere, in mentioning this sensational 'discovery,' it was stated—

'If confirmed, it proves that meteorites are not mere wandering fragments of an exploded sun, but they have formed part of a world furnished with water and organized life.' 'It is the opinion,' it was added, 'of many savants that meteorites do not belong to the solar system, but are visitors from far distant spheres;' and it would 'seem to prove that round these suns revolve worlds with water and atmosphere, and an organized life almost identical with our own.'[1]

Hartwig also devotes an interesting chapter of his *Aerial World*[2] to aerolites, mentioning the various

---

[1] Newspaper paragraph, *Evening Standard*, May 1882.
[2] Hartwig's *Aerial World*, p. 350.

opinions which have been entertained regarding their origin. Some supposing them to be the product of our own volcanoes, others ascribing them to the Moon, others to coming from the depths of space, others to comets.

Skertchly[1] advocates a solar origin for at least some of them, and for a stellar origin for others.

As it would not be possible for our own volcanoes to hurl the meteoric stones to a point beyond the gravitation power of Earth, so we may put aside the thought that they are the product of volcanoes of other planets, either of our own or of stellar systems. And it would seem a wild view to suppose even that stones of a few pounds weight, or a few tons weight, have been ejected from any other sun beyond the power of their attracting gravity, looking to the tremendous gulf which exists between our Sun and the nearest of the fixed stars. But whether possible or not, one thing is clear, that if such meteoric stones have been projected from either our Sun or from any other sun, they could bring with them no organic life, either vegetable or animal, because, as we have already seen, the Sun and stars are in a condition in which no such life could exist.

Unfortunately for the grave discovery, it was speedily demolished by Carl Vogt. 'The organisms announced by M. Hahn have no existence; what has been described and drawn as such, results from crystalline conformations which are absolutely inorganic.'[2]

The antiquated thought that the asteroids or that meteoric stones are pieces of an exploded planet may be dismissed as altogether unwarranted.

But by means of the investigations of the Italian

[1] Skertchly, p. 150.   [2] *Knowledge*, i. 251.

astronomer Schiaparelli, and of Adams and Leverrier, coincidence has recently been found between the meteoric system and comet paths, so that apparently some of the meteoric showers are derived from cometary tails. The subject has been fully treated of in Guillemin's *World of Comets*,[1] and it is not necessary to the present question to enter upon it here, but the fact naturally leads to our next looking at the case of

2. THE COMETS.

Kepler considered 'comets to be huge, uncommon creatures, generated in the celestial spaces, and that they were made to the end that the ethereal fluid might not be more void of monsters than the ocean is of whales and other great thieving fishes.'[2]

But the notions regarding their appearance have been equally strange.

'The extraordinary aspect of comets,' says Sir John Herschel, 'their rapid and seemingly irregular motions, the unexpected manner in which they often burst upon us, and the imposing magnitudes which they occasionally assume, have in all ages rendered them objects of astonishment, not unmixed with superstitious dread to the uninstructed, and an enigma to those most conversant with the wonders of creation and the operations of natural causes.'[3]

The superstition has principally taken the form of imagining them to presage calamitous coming events, such as the deaths of eminent personages. Nor has the superstition been confined to the unlearned. Kings

---

[1] Guillemin's *World of Comets*, chap. xii. pp. 417–451. See Mrs. Somerville's *Connection of the Physical Sciences*, 10th ed. p. 376.

[2] Milne's *Essay*, p. 43.   [3] *Outlines of Astronomy*, § 554.

and priests have mingled with the crowd. Pope Calixtus, believing the appearance of the splendid comet of 1456

'to be at once the sign and instrument of divine wrath, was so frightened at its appearance that he ordered public prayers to be offered up in every town, and the bells to be tolled at noon of each day to warn the people to supplicate the mercy of Heaven. He at the same time excommunicated both the comet and the Turks, whose arms had lately proved victorious over the Christians'![1]

Indeed, if a comet were not in its entirety a messenger of evil, what was to hinder a part of it to be; for we are told by one writer 'that if the comet (of 1680) is not a visible sign of the anger of God, the tail may be.'[2] The myths and superstitions which have arisen regarding them have been many, and will be found mentioned in various books.[3] Some people have from unknown sources obtained wonderful information regarding them. Thus we learn that Bodin, a French lawyer (1575), thought with Democritus 'that comets are spirits which, having lived on the Earth innumerable ages, and having at last completed their term of existence, celebrate their last triumphs, or are recalled to heaven in the form of shining stars,'[4] so that their absence from Earth by their absorption or conversion into comets was, it appears, the reason of famines, epidemics, and civil wars; nay, it would seem that the apparition of this cometary spirit in one case, sad to say, occasioned *all the cats in Westphalia to be ill.*[5]

---

[1] Dick's *Sidereal Heavens*, p. 332.

[2] Flammarion's *Marvels*, p. 351.

[3] See among others Flammarion's *Astronomical Myths*, also his *Marvels of the Heavens;* Milne's *Prize Essay on Comets* (1828); Guillemin's *World of Comets;* *The Comet* (Biela's), by Arago (1832).

[4] Milne's *Essay*, p. 43.   [5] Arago, *The Comet*, p. 77.

But the dread has extended somewhat more legitimately to the fear lest a comet should tilt against the Earth; and it has even been thought that the great deluge was produced by an awkward collision of this kind. Such fear arose in 1773 from the apprehension that a comet would cross the Earth's orbit, that 'persons of weak minds died of fright, and women miscarried. There were not wanting people who knew too well the art of turning to their advantage the alarm inspired by the approaching comet, *and places in Paradise were sold at a very high rate.*'[1] In 1832, therefore, to allay similar apprehensions arising out of the expected proximity of Biela's comet that year, M. Arago wrote his little treatise on *The Comet*,[2] in which he pointed out, *inter alia*, that collision would not occur, and gave the comforting assurance that the calculations of probabilities show that of 281,000,000 chances there is only one unfavourable. However, this one chance is quite possible to happen in our own day, and it has been thought that the tails of the comets of 1843[3] and 1861[4] did graze the Earth, but without producing any sensible effect. The direct charge of (for a comet) a *solid* nucleus might possibly be another matter.

But our sole interest at present arises upon the question of habitability, and notwithstanding the marked differences which exist between the planets and the comets, the latter have not been without advocates for their being peopled, and that, too, by a race of intelligent beings even superior to ourselves.

Fontenelle's Marchioness is probably responsible for

---

[1] Dick's *Sidereal Heavens*, p. 314.   [2] Arago's *Comet*, p. 62.
[3] Somerville, p. 369.
[4] Guillemin's *Heavens*, p. 260, and other authorities.

suggesting the fanciful idea upon which the supposition has been based. In the fifth evening, the conversation having turned upon our present subject of discourse, she observed—

'There can be nothing so diverting as to change vortices. We that never go out of our own sphere lead but a dull life; if the inhabitants of a comet had but the wit to foresee the time when they are to come into our world, they who had already made the voyage could tell their neighbours beforehand what they would see, and could inform them that they would discover a planet with a great ring about it, meaning our Saturn; they would also say you shall see another planet, which has four little ones to wait on it; and perhaps some of them, resolving to observe the very moment of their entrance into our world, would presently cry out, A new sun, a new sun, as sailors used to cry, Land, land.'[1]

Lambert improved on this idea by figuring to himself the comets as travelling globes peopled with astronomers,[2] carried from world to world to form enlarged conceptions, and thus apostrophizes the inhabitants as

'happy intelligences, how excellent must be the frame of your nature! Myriads of ages pass away with you like so many days with the inhabitants of the Earth. Our largest measurements are your infinitely small quantities,' and so forth.[3]

Mr. Milne,[4] following in the same line, contended for

---

[1] Fontenelle's *Plurality of Worlds*, p. 141.
[2] Dick's *Sidereal Heavens*, p. 342.
[3] Perhaps some troublesome objector might curiously desiderate information regarding the *locale* of the mines out of which these happy intelligences would dig material for a twenty-five inch refractor, or inconveniently ask upon which end of a comet's tail would secure foundation be had for building an equatorial observatory, of largest human measurement.
[4] Milne's *Essay*, p. 139.

the habitability of those comets 'whose advanced state of maturity renders the Sun's influence incapable of materially affecting the surface of the nucleus;' and, ingeniously reasoning against the objections to the supposition, and disclaiming the necessity for assuming the inhabitants to have bodily frames resembling those of terrestrial beings, he would conclude (p. 144) that

'if we estimate the intelligence of beings by the knowledge which their place in the Universe is fitted to impart, we are compelled to regard the cometary inhabitants as of an order even superior to the creatures of the Earth. When, for example, they find themselves passing through the midst of the satellites, those small bodies which we can scarcely discern with telescopes, or when they are brought so close to the planet Saturn that they can examine the wonderful phenomenon of his rings even with the naked eye, or when at the perihelion passage they are able to observe everything on the surface of the Sun, that great luminary, the mysterious source of life, and light, and energy to the system, what spectacles of delightful contemplation must they enjoy, and what means of attaining an acquaintance with the works of nature, infinitely greater than any which we shall ever command! Traversing as they do the whole extent of that system of which the Earth forms so insignificant a member, and directing their course far beyond its known limits into those regions of space whose dark and unfathomable nature it will for ever baffle human penetration to explore, the beings who have their abodes on comets must be familiar with many important truths of which we can obtain only a few casual glimpses, and witness such glorious and sublime displays of the manifold wonders of creation as must afford to them the noblest conceptions of that Almighty Being by whose wisdom they were constructed, and by whose power they are still sustained.'

Dr. Dick[1]—after referring to Mr. Whiston's rather different supposition, that the comets were, from their exposure to the extremes of heat and cold, the places of punishment for the wicked, but considering the boundless beneficence of the Divine Being, holding it was impossible to suppose so vast a number of these bodies would be created for such an end—concludes (conformably to his strong leanings to the habitability of all Heavenly Bodies) a long discourse on comets by saying:—

'What I conceive to be one of the main designs of the Creator in the formation of such a vast number of splendid bodies is, that they may serve as habitations for myriads of intellectual beings, to whom the Almighty displays His perfections in a peculiar manner, and on whom He bestows the riches of His beneficence.' 'We ought to contemplate the approach of a comet, not as an object of terror or a harbinger of evil, but as a splendid world, of a different construction from ours, conveying millions of happy beings to survey a new region of the divine empire, to contemplate new scenes of creating power, and to celebrate in loftier strains the wonders of omnipotence. Viewing the comets in this light, what an immense population must be contained within the limits of the solar system!'[2]

Truly, were this assumption well founded, the concluding exclamation would be just. For although Dr. Dick seems to limit his thoughts to comets having a solid nucleus, the number of comets of all kinds which pervade or invade the solar system is great, and while figures differ with opinions and modes of calculation,[3] yet all are agreed that they are to be counted by millions,

[1] *Sidereal Heavens*, p. 338.   [2] *Ibid.* p. 356.
[3] See Dick's *Sidereal Heavens*, p. 350; Somerville, p. 377; and particularly Guillemin's *Comets*, chap. vi. § 1.

and even, some say, by billions; for, upon one estimate, Guillemin reckons the number of comets traversing the solar system within the orbit of Neptune at not less than 45,500,000, while, taking in all which have gravitated at least once round the Sun, he arrives at a minimum of 74,000 billions.

The thought upon which the supposition rests is, however, rather to be regarded as a fanciful conceit of a philosophical imagination than as having any substantial groundwork, and a single observation may suffice to put it aside. While we have a few comets of short period, the average duration of the cometary orbit has been said to be 1000 years,[1] but many have periods of enormous length. Arago[2] specifies a number, ranging up to 100,000 years; while a list made up by Guillemin[3] gives for the two longest respectively 1,840,000 and 2,800,000 years. Now the portions of these periods during which the two comets pay flying visits to the interior parts of the solar system form but small fractions of the whole; by far the larger part of their existence is spent in distant regions, and, in the case of many, thousands of cold years are occupied in slowly traversing the dark realms of vacant space. Thus Donati's comet,[4] with a period of 2100 years, wanders out eighty times the distance of Neptune. 'The comet of 1845,' says Guillemin,[5] 'recedes to a distance from the Sun of two and a half times the mean distance of the Sun from the Earth. The voyage outward, it is true, takes 1,400,000 years, and the return also 1,400,000 years.'

What 'spectacles of delightful contemplation,' may it

---

[1] Dick, p. 323.　　　[2] Arago's *Popular Astronomy*, vol. i. p. 572.
[3] Guillemin's *Comets*, p. 145.
[4] Herschel's *Popular Lectures*, p. 139.　　　[5] *Comets*, p. 152.

be asked, could the inhabitants have to observe,—what 'enlarged conceptions' could they receive during even two millions of the long years thus spent? They would be endured in blackest of cheerless darkness, without hailing a solitary planet; and even other comets, supposing them possessed, like the luminous fishes of the great sea depths, of inherent light capable of rendering them dimly visible, if at all, would cross their paths at tremendous intervals of time in that unfathomable ocean of space, nay, perhaps in some voyages never. Then with what length of life would it be needful to endow them? or would thousands of generations die and be buried in some flickering ray or gassy clod without ever having witnessed the glorious sights for the enjoyment of which these 'superior' persons were, according to the theory, created? Nay, rather let us suppose, were comets inhabited, the approach to such unwonted sights would only inspire the community with liveliest terror; and when passing from the chilly regions of sunless void they dashed into the blinding presence of the great central solar fire, these 'happy intelligences' would experience doubtless a 'familiarity with important truths;' but a proximate familiarity of a kind which, if it did not at once utterly destroy them, would leave burning impressions, in the midst of which, dancing with agony round the nucleus till frizzled up, with what patience could their distracted minds regard the suggestion of tranquil astronomical investigations! The chimerical idea must therefore be dismissed, while there are other and at least equally conclusive grounds on which the habitability may be rejected.

To enter fully upon the subject would, however, be to occupy disproportionate space. It can only be treated briefly.

Guillemin describes the difference between a planet and a comet in these words:—

'A globular mass, solid or liquid, surrounded on all sides by a light and comparatively thin aeriform envelope, is, perhaps, from a physical point of view, the simplest description of a planet. The comparative stability is due, on the one hand, to the preponderance of the central globe, where general phenomena are modified only at long intervals; and, on the other, to the trifling depth of the atmosphere, the portion of the planet the most subject to variation and internal change. In comets, we have seen, this relation is reversed, and the atmosphere or nebulous envelope constitutes the entire body, or at all events greatly preponderates. At the utmost, we can only conjecture that in some comets the nucleus is solid or liquid. Certainly its volume is generally but a very insignificant portion of the entire nebulosity, even if we except the tail.'[1]

There are, however, comets without a head or nucleus; others without a tail, others with an occasional tail. But 'the distinctive sign of a comet' is to be found in 'its orbit, its large eccentricity, great inclination, direction,' etc.[2]

1. In this, then, we discover the first element which destroys resemblance to our planet. Comets[3] traverse the starry vault and fly towards the Sun from every direction, and unlike the planets, which all move in a similar plane and from west to east, many move from east to west; and calculations made seem to show that as many are 'retrograde' as 'direct.'[4] But, in addition, their course is sometimes elliptical, sometimes parabolical,

---
[1] *World of Comets*, p. 247. [2] *Ibid.* p. 197.
[3] Guillemin's *Heavens*, p. 241.
[4] Dick's *Sidereal Heavens*, p. 351; Guillemin's *Comets*, p. 188.

and although there has been question about it,[1] yet lists have been made [2] of some considered to be hyperbolical —that is to say, comets whose eccentricity of orbit is such as to show us they entered our system for the first time, and that having once visited it, they can never return. These are circumstances which would indicate, upon the Nebular Hypothesis, that they are not the offspring of the Sun, and therefore possess *no community of origin with its planets*. Some have supposed that they have come from other systems, or have been ejected from other stars; and Mr. Proctor, in propounding or supporting this view, observes that, supposing projection from the nearest star, 'eight million years would be the shortest time in which any comet could traverse the space separating our system from it.'[3] But whether coming from other systems or not, this fact creates a material and important difference from our world, and indeed from every other orb. Laplace regarded them 'as small nebulæ wandering from one solar system to another, and formed by the condensation of the nebulous matter scattered with such profusion throughout the universe.'[4] And if I might venture to express an opinion, this is a view which seems possessed of much greater probability than that they are bodies ejected from the Sun or from any star, whether near or remote.

2. The next peculiarity is the excessive changes to which comets are subjected.

(1.) There is, first of all, the enormous difference between their velocity at aphelion and at perihelion. Thus the comet of 1680 was estimated by Sir Isaac

---

[1] Herschel's *Outlines*, § 564, p. 378.   [2] *World of Comets*, p. 168.
[3] *Expanse of Heaven*, p. 149.
[4] Quoted without reference in *World of Comets*, p. 169.

Newton to move at its perihelion at the rate of 880,000 miles per hour;[1] a Mr. Squire, by two different calculations, making it out to be no less than 1,240,000 miles per hour. While its motion when at aphelion, when it is 13,000 millions of miles distant from the Sun, or seven times the distance of Uranus, is only 3 yards a second, or little more than 6 miles in an hour.[2]

(2.) Comets are subject to sudden changes of brilliancy in the light of their tails; see this described by Guillemin, *World of Comets*, pp. 305–308.

(3.) As a comet approaches or retires from the Sun, and owing in one way or another to the Sun's action, its size alters rapidly. 'It is on the nucleus,' says Milne, 'that the action of the Sun's heat produces the most remarkable effects, bringing about physical changes as sudden in their occurrence as they are often difficult of explanation. Of this nature were the phenomena observed by Schröeter in the comet of 1799 during the month of its greatest proximity to the Sun.'[3] And he mentions the changes which embraced a reduction in diameter of the nucleus in two days to two-thirds, and in another day the nebulous matter had lessened one fourth part. Guillemin, dealing with the same subject, takes as an example Donati's comet (1858).[4] The diameter of its nucleus on 19th July was 4660 miles; on 5th October, having gradually decreased, it was only 400 miles, next day it attained 800 miles, and on 8th October to 1120 miles; on 10th was again reduced to 630 miles.

While the head of the comet, or rather the nebulous covering of the nucleus, is thus suffering vast change, the production or prolongation of the tail, or its con-

[1] Dick, p. 308.
[2] Guillemin's *Heavens*, p. 252.
[3] Milne's *Essay*, p. 27.
[4] *World of Comets*, p. 240.

traction, proceeds with inconceivable rapidity, and stretches out often to enormous distances, where just previously it was invisible. The tail of the comet of 1811 attained the length of 109,400,000 miles, that of 1680 to 149,000,000 miles, that of 1843 to 198,800,000 miles.[1] Nor is this all, for the tail issues on its approach to the Sun from nil, and upon retreating from the Sun returns to nil — at least visibly. Thus, in the case of Donati's comet (1858),[2] the first appearance of the tail was observed on 14th August, 46 days before its perihelion passage, and it had then an apparent length of but ten minutes. By the end of August it attained two degrees; and, 10th October, 11 days after its perihelion passage, its maximum of 64 degrees rapidly began to decrease, till on 3rd December it was reduced to only 55 minutes.

(4.) Comets are also subject to another species of disturbance or change of form as they approach the Sun, in the emanation from the nucleus of aigrettes or triangular eruptions, or jets or streams of light from the part turned towards the Sun.[3] In an interval of seventeen days, M. Chacornac was able to distinguish in the comet of 1862 'the formation of thirteen of these jets, similar to jets of steam, and alternately directed towards the Sun and to the east of it, that is to say, in a direction opposite to the movement of the comet; after each of these emissions the nebulous matter accumulated at the end of the jet seemed driven back by a repulsive force emanating from the Sun, and then flowed in the direction of the tail.'[4]

(5.) Not only is the cometary tail sometimes developed in two or more branches, but what, supposing

---

[1] See list in *World of Comets*, p. 222.   [2] *Ibid.* p. 227.
[3] Herschel's *Outlines*, § 569.   [4] Guillemin's *Heavens*, p. 259.

comets to be inhabited, would be a terrifying catastrophe, one comet—Biela's—was observed to divide into two distinct comets, which, after thus parting company, continued to journey on together at a separating distance, at first, of about 180,000 miles,—considerably less than the distance of the Moon from us,—but afterwards much increased; and the two portions appear not again to have united.[1] It was probably not a solitary instance among comets of what would to inhabitants be such a dread disaster.

3. It is principally, however, upon the fact of the vast extremes of temperature to which comets are subjected, that opinion against the habitability is generally rested. We have seen to what immense distances in space away from the Sun some comets reach. But without taking such extreme cases, observe one, 'the comet of short period,' that of Encke, which at perihelion is 31 millions of miles from the Sun, or within the orbit of Mercury; and at aphelion is 375,000,000 miles, or near the orbit of Jupiter.[2] Halley's, which is of 76 years' period, approaches the Sun to 54,000,000 miles, or within the orbit of Venus, and removes from it to a distance at aphelion of 3,100,000,000 miles, or 250 millions of miles beyond the orbit of Neptune. Upon this comet Guillemin observes:—

'These enormous variations in distance would lead us to suppose most astonishing differences in the quantity of light and heat received by the comet from the Sun; and, in fact, the intensity of these physical agents varies in the ratio of 3000 to 1, or, as it may be put, the Sun's

---

[1] See Herschel's *Outlines*, § 580; also a pretty full account in Guillemin's *Comets*, chap. viii. § 3, pp. 258-267.

[2] Guillemin's *Heavens*, p. 244.

light and heat arrive at the comet with a force 3000 times more considerable at perihelion than at aphelion.'[1]

The comet of 1680,[2] whose period[3] has been estimated at 3000 years, is at aphelion 853 times the distance of the Earth from the Sun (or five-and-twenty times farther than Halley's comet),[4] approached at perihelion about 146,000 miles from the surface or photosphere of the Sun—passing thus deep down through the corona; and was estimated by Newton to have been subjected to an intensity of heat 2000 times that of red-hot iron—'a term of comparison,' Sir John Herschel says, 'indeed of a very vague description, and which modern thermotics do not recognise as affording a legitimate measure of radiant heat.'[5]

But this approach was far exceeded by the wonderful comet of 1843, which, with a period of 8813 years,[6] entered the corona at a distance of only 30,000 miles

---

[1] Guillemin's *Heavens*, p. 243.

[2] The superstitious panic which was excited by this comet was the cause of publication by the French philosopher Pierre Bayle of his work, called *Miscellaneous Reflections occasioned by the Comet which appeared in 1680, chiefly tending to explode Popular Superstitions : written to a Doctor of the Sorbon*. He seriously argued with special reference to historical events against the mistake—'a remnant of pagan superstition'—of supposing that the comets were the presage of evils to the Earth, which, if they were, would be a miracle — an idea not to be approved. His view, sound and rational (Translation, London 1708, vol. i. p. 111), was that 'we must believe that comets are the ordinary works of nature, which, without regard to the happiness or misery of mankind, are transported from one part of the Heavens to another by virtue of the general laws of motion;' but his book—a curious exhibition of the kind of thoughts afloat at that time—is of so rambling a character, that it hardly needed he should ingenuously tell us (vol. ii. p. 535), 'when I first sat down to write I did not know what I should say at the third page, and almost all I have said occurred as I wrote to you, without my ever having had a thought on't before in the whole course of my life'!

[3] Flammarion's *Marvels*, p. 197. [4] Dick's *Sidereal Heavens*, p. 308.
[5] *Outlines*, § 593. [6] Guillemin's *Comets*, p. 145.

from the Sun's photosphere, the effect of which Sir John Herschel estimates to be equivalent to an exposure to the glare of light and heat of '47,000 suns such as we experience the warmth of, on the materials of which the Earth's surface is composed.' And he adds:—

'To form some practical idea of it, we may compare it with what is recorded of Parker's great lens, whose diameter was 32¼ inches, and focal length, six feet eight inches.' 'The heat to which the comet was subjected surpassed that in the focus of the lens in question, on the lowest calculation, in the proportion of 24¼ to 1 without, or 3½ to 1 with the concentrating lens. Yet that lens so used melted carnelian, agate, and rock crystal!'[1]

But to this degree of heat, he explains, the comet would be exposed only for little more than two hours; as it receded, the Sun's potency would, of course, gradually diminish, although, as comets generally shine more brightly after passing the perihelion, much of that heat would necessarily be retained for a period of time.

A body which could survive such a tremendous exposure must be of a different constitution altogether from that of our world, and this of itself destroys all analogy to the Earth, and with it all claim to habitability. Mr. Milne,[2] however, ingeniously argues that there is a certain point in bodies beyond which, whatever may be the means employed, their temperature never can be elevated, and instances water, which can only be heated up to 212°; 'but all the heat which we employ in the endeavour to raise this temperature higher is only dissipated in the ensuing evaporation.' True, but this means that, exposed to heat beyond a certain point, *dissolution happens;* and he forgets that the question

[1] Herschel's *Outlines*, § 592. [2] Milne's *Essay*, p. 140.

is not about the comet itself, but as to the capacity of beings possessing life to endure more than a certain degree of heat, and with the possibility of vegetation, upon which animal life is dependent, enduring a scorching temperature of fiery energy. It seems enough to state the fact of such heat and—especially after what has been already said in reference to the planet Mars—of such violent extremes of heat and cold as are implied in cometary existence, to exclude the thought of their habitability.[1]

4. Still it is well to exhaust the subject, for there is yet another element of consideration in the nature or constitution of comets.

(1.) And, in the first place, they are of the lightest or most attenuated material. Flammarion forcibly expresses himself thus:—

'Their rarity is such that out of the tails of certain comets we should be able to cut a piece the size of Notre Dame, and inhale it as a homœopathic inspiration. Comets have been seen several million leagues long, whose mass was nevertheless so small that it would have been possible, without fatigue, to carry it on one's shoulder.'[2]

Sir John Herschel is even more emphatic; for referring to a calculation of Newton's, he observes the tail of a great comet, 'for aught we can tell, may consist of only a few pounds or even ounces of matter.'[3] So small, indeed, are the cometary masses, that they exercise no

---

[1] There are doubtless comets whose approach to the Sun may not be sufficiently near to entail great heat; but this would mean the total, or nearly total, exposure to incessant inconceivable frigidity, which would be just as fatal.
[2] Flammarion's *Marvels*, p. 209.   [3] *Outlines*, § 559, footnote.

appreciable influence on the planets they approach. That of 1770 passed within about six times the distance of the Moon from us without affecting tides,[1] and the same comet in 1779 passed near the moons of Jupiter without disturbing them.[2]

(2.) They can be possessed of no consistency or solidity, seeing they are subject to such violent changes as have been above mentioned. Indeed, Professor Tait has suggested, although his suggestion has not been admitted, that comets consist of a mass of meteorites,[3] or a cloud of small stones; but if they do, the meteorites must be smaller than dust, and infinitely scattered or diffused. In any view, so little solidity do they possess, that stars are seen through the tails, and even through some nuclei, which are in general more dense than the coma which envelopes them.

(3.) But the spectroscope seems to disclose that the nuclei of comets are partly self-luminous, and are partly reflective of light from the Sun. There may be in them some amount of incandescent solidity, but the coma which surrounds and is derived from the nucleus, and also forms the tail, is gaseous and incandescent. Schellen deduces from the investigations made that the nucleus 'is very possibly composed of glowing gas containing carbon.'[4] Guillemin observes that

'in a chemical point of view the comets—few in number, it is true—which have as yet been subjected to examination are of very simple constitution. They consist of simple carbon, or of a compound of carbon and hydrogen,' according to the comparisons made by Mr. Huggins; carbonic oxide or carbonic acid, according to the re-

---

[1] Somerville, p. 358.   [2] Guillemin's *Comets*, p. 278.
[3] Tait's *Recent Advances*, p. 255.
[4] Schellen's *Spectrum Analysis*, p. 578.

searches of Father Secchi. The Italian astronomer was therefore justified in saying, "It is very remarkable that all the comets observed up to the present time have the bands of carbon."[1]

The spectroscopic investigations regarding comets will be found mentioned in Schellen's *Spectrum Analysis*, pp. 568-581, and in Guillemin's *World of Comets*, pp. 315-327, up to their respective dates; but looking to the fact thus elicited of the gaseous nature of the coma, and of the presumably deadly character of the gaseous ingredients, and to all the circumstances previously stated, each of which seems sufficient of itself to prove fatal, we cannot but conclude that the romantic notions once held by some regarding comets are to be altogether put aside, and that these multitudinous bodies must be declared uninhabitable.[2]

---

[1] Guillemin's *Comets*, p. 327.

[2] Indeed there can be little fellowship between us and those light-headed Arabs of desert space, who keep such irregular hours, wear such disorderly garb, and employ themselves in rushing wildly out of their dark retreats to terrify timid old maids. We can now see what praise was due to that brave and pious pontiff, Calixtus, who so benevolently exerted his vast power to excommunicate one of those audacious intruders. But while engaged in a work so meritoriously useful, why, may we ask, did he not proceed a step further, and lay the whole kingdom of comets under the potent ban of a papal interdict? What an unknown amount of earthly disaster might he not then have averted!

VIII.

PASSING from the smallest or lightest of all the bodies of our system, but bodies to be counted as we have seen by millions or billions, we now arrive at length at an examination of the four great outer planets, and it will be found that the question under discussion falls to be determined with regard to them by different classes of facts from those which have decided the state or the fate of the minor planets. The first and most important is—

JUPITER.

M. Flammarion describes the planet Jupiter as

'a charming one, so far at least as we are able to judge from afar, and without going there. To begin with, a continual spring rejoices its surface. If it is ornamented with flowers, which we do not doubt, though we know not of what these flowers consist, they do not only survive "the span of a morning" as our roses do, but live much longer. Scarcely have the oldest begun to dry up and fade, but they are replaced by lovely buds opening before the first have died away.'[1]

This is very poetical, and may create a longing to reach that favoured but distant land. Yet, as we shall see, it is about as well founded as the views of Sir David Brewster, who, in endeavouring to combat and remove the 'objections,' or rather 'difficulties,' regarding this planet, of 'persons who have only a superficial know-

[1] Flammarion's *Marvels*, p. 174.

ledge of astronomy, though firmly believing in a plurality of worlds,'[1] advances speculative hints which do not argue that he had applied his great scientific knowledge to good purpose, when his creed as a philosopher and his hope as a Christian seemed to be assailed and endangered.

We shall find as we proceed that extravagances which have been broached by this and that person must vanish before the light of day. Jovian giants of fourteen feet high, alike with humbler pigmies of two; bat-winged creatures flapping about among the belted clouds; bottomless waters filled with cartilaginous and glutinous monsters — boneless, watery, and pulpy; Polyphemuses with one eye-ball, or Arguses with a hundred, stalking along upon a land of pumice-stone, or tabasheer, or spongy platina; or beings of another mould, having their homes in subterranean cities warmed by central fires, or in crystal caves cooled by ocean tides, or floating with the Nereids upon the deep, or mounting upon wings as eagles, or with the lighter pinions of the dove, that they may flee away and be at rest. Alas! these whimsical fancies, how pleasing soever they be in a picture, to the pluralist, can, in the face of a perverse reality, find no place in serious judgment. Jupiter is, even more hopelessly than Mars, no dwelling-place for man or beast, or even, Sir David notwithstanding, any 'faultless monster whom the world ne'er saw.'

Putting aside the literature of Jovian monsters, and without going minutely into the matter, these well-known facts may be stated, keeping in mind that figures differ considerably in different books, and that precise exactitude in this respect is not here material.

[1] *More Worlds than One*, p. 68.

Jupiter, then, is by far the largest of all the planets. Its diameter, variously estimated, may be stated to be about 86,000 miles, or eleven times the diameter of the Earth. In volume it is 1300 times or thereby, and in mass 340 times the size of the Earth. In density it is only about one-fourth of the density of the Earth, or little more than the density of water. The force of gravity at its surface is $2\frac{1}{2}$ times that at the surface of the Earth. Its axis is almost perpendicular to its orbit, depriving it of change of seasons. It revolves on its axis in its day of ten hours. Its distance from the Sun is say 490 millions of miles, or five times that of the Earth. The light and heat it receives from the Sun is about 1/25th or 1/27th that received from the Sun by the Earth, and it has four moons of differing size, but from which the heat and light it receives may be said to be next to nil.

Now these are facts which do not bear, in the question of plurality, in the same direct way upon the habitable capacity of Jupiter as we have seen they do upon that of the planets already considered. People have puzzled themselves as to the crushing effect of the greater power of gravity on life and vegetation; but, as we shall see, neither life nor vegetation can exist on Jupiter, and the puzzle may therefore be spared. The planet is uninhabitable, but not because the Sun is so distant as to deprive of heat sufficient to preserve or sustain life. Therefore we need not torture our imagination, as some have done, to find in cloud or otherwise shelter and covering for ideal limbs. The giant planet, as we shall see, wants not heat. It is not like Mars, frost-bound at the poles either in winter or at any other time. Nay, it has no winter. It has no change of seasons. Spring perpetual reigns. But neither is the absence of changing seasons the cause of failure of either vegetable or animal life. In other circumstances—had the planet resembled in condition those planets which lie within its orbit—the facts which have been stated would have sufficed of themselves to negative life. It is in the condition of Jupiter —a condition not peculiar to it alone, but held to a certain extent, if not altogether, in common with Saturn, Uranus, and Neptune—we have to seek the elements of deduction. And the peculiarity so to be sought for is to be found in the atmosphere in which it is enwrapt, or of which it consists—the belting of cloud by which it is surrounded.

We are indebted to Mr. Proctor for the views which have been suggested, and now, I believe, are generally admitted, by which Jupiter must be untenanted even of cartilaginous monsters. They amount to this, that the giant planet is glowing still with nebulous heat.

Mr. Proctor tells us in one of his papers on the subject, that he had been led in 1869 [1] to present a new theory respecting Jupiter's condition when he was visiting 'other worlds than ours.' Subsequently 'observations had been made which placed the new theory on a somewhat firm basis.' His views are scattered through several of his works; I have extracted them from six, and as he is a most voluminous writer, there may possibly be others dealing with the subject which I have not seen. Necessarily there is a little reiteration, and sometimes expansion and variation. Referring to Mr. Proctor's works themselves for fuller statement, I shall endeavour to combine and reduce his observations

[1] *Flowers of the Sky*, p. 191.

to orderly statement, and in this view the first point which arises is to ascertain the appearance of the cloud belts, and any phenomena connected therewith. They are thus described in *Other Worlds than Ours*:—

'The belts of Jupiter are commonly arranged with a certain symmetry on either side of the great equatorial bright belt, but sometimes there is a rather marked contrast between the northern and the southern halves of the planet. In colour the dark belts are usually, when seen with suitable telescopic power, of a coppery, ruddy, or even purplish tint, while the intermediate light bands vary from a pearly white in the equatorial belt, through yellowish white in the middle latitudes of both hemispheres, to a greyish or even bluish tint at the poles.'[1]

The arrangement of the belts is thus described:—

'There is commonly a bright belt across the middle of the disc, which goes by the name of the equatorial belt.' 'On either side of this belt there are commonly two dark belts.' 'Then usually follow several alternate light and dark streaks up to the polar regions.'[2]

But the belts do not maintain by any means a uniform appearance. For they

'vary greatly in form, extent, and general appearance. At one time the dusky belts cover a large proportion of the planet's disc, at another they are singularly narrow. Now they are very regularly disposed, now they seem in some way under the action of disturbing forces of great intensity, causing them to assume the most irregular figure.' 'The number of belts is singularly variable. Sometimes only one has been seen, at others there have been as many as five or six on each side of the planet's equator. In the course of a single hour Cassini saw a complete new belt form on the planet; and on 13th December 1690, two well-marked belts vanished com-

---

[1] P. 129. See also *Orbs around us*, pp. 138, 139.
[2] *Orbs around us*, p. 128.

pletely, while a third had almost disappeared in the same short interval of time.'[1]

Elsewhere Mr. Proctor says :—

'Mighty masses of cloud, such as would suffice to enwrap the entire globe on which we live, form over large regions of Jupiter or Saturn, change rapidly in shape, and vanish in the course of a few minutes.'[2]

Equally changeful are the colours of the belts. The equatorial belt

'is usually of a pearly white tint, and has long been recognised as one of the most constant features of the planet's aspect. As the mean surface of this belt cannot be less than a fifth of the whole surface of the planet, it is clear that any changes which may take place in its general aspect cannot but be of the utmost significance. Now, during the autumn of 1869 and the spring of 1870, this belt has been more strongly coloured than any part of the planet. Mr. Browning observing Jupiter in the earlier part of the above-named interval, found the equatorial belt of a greenish yellow colour, which deepened in October 1869 to a full ochreish yellow, and in January of the present year had assumed an even darker tint, resembling yellow ochre. On one occasion, and on one only, he detected this tint in the first bright belt north of the equator. While thus exhibiting strongly marked and changing colours, the equatorial belt has lost its right to be called *par excellence* the bright belt of the planet, being considerably inferior in brilliancy to the narrow bright belts north and south of it.'[3]

It appears, however, from a paper by Mr. Proctor in 1873, that the planet had

'gradually resumed its normal aspect after four or five years.' 'The zone is now of a creamy white colour, its ordinary hue.'[4]

[1] *Other Worlds*, p. 132.   [2] *Familiar Science Studies*, p. 60.
[3] *Other Worlds*, p. 135.   [4] *Our Place among Infinities*, p. 109.

But there are some phenomena important to be noticed, and the first of these is the existence of spots resembling those upon the Sun:—

'These spots,' Mr. Proctor says, 'have been described as black, though surely their appearing of that hue must be ascribed to the effect of contrast. Now these dark spots which have been seen by Cassini, Mädler, Schwabe, Airy, and others, may be regarded as the real surface of the planet (unless they belong to a yet deeper cloud layer), seen for a while through openings in the cloud bed to which the dusky belts belong.'[1]

A conjecture which apparently was afterwards abandoned by Mr. Proctor, for in 1882 he wrote 'that if Jupiter has any nucleus at all, it is not visible to us.'[2]

The spots would appear, like those of the Sun, to be subject to rapid change. We are told of one witnessed by South on 3d June 1839, which he estimated at 20,000 miles in diameter, but which thirty-four minutes after observation commenced had only left three miserable scraps.[3]

Another curious phenomenon is that of the appearance of 'white spots—some small, some large—which are seen to form from time to time along the chief belts in certain cases with singular regularity,' 'set side by side like rows of eggs upon a string.'[4] Mr. Brett, describing these 'egg-shaped clouds,' tells us

'that they cast shadows; that is to say, the light patches are bounded on the side farthest from the Sun by a dark border, shaded off softly towards the light, and showing

---

[1] *Other Worlds*, p. 131.
[2] *Familiar Science Studies*, p. 69. See also *Flowers of the Sky*, p. 204.
[3] *Flowers of the Sky*, p. 204. [4] *Familiar Science Studies*, pp. 65, 66.

in a distinct manner that the patches are projected or relieved from the body of the planet.'[1]

The larger white globular spots, it seems, represent masses of cloud 5000 or 6000 miles in diameter.

But still another noticeable fact is to be found in the streaks which cross the belts.

One of the most remarkable of these streaks was observed by Mr. Long of Manchester on 29th February 1860. Its length was

'about 10,000 miles, and its width at the least 500 miles, so that its superficial extent was much greater than the whole area of Europe. But wonderful as this rift appears when thus regarded, its mere dimensions and its singular position were by no means the most remarkable features it presented. First of all, it remained as a rift certainly until April 10, or for six weeks, and probably much longer. It passed away to the dark side of Jupiter, to return again after the Jovian night to the illuminated hemisphere, during at least a hundred Jovian days; and assuredly nothing in the behaviour of terrestrial clouds affords any analogue of the remarkable fact. The arrangement of our clouds depends far more directly on the succession of day and night than that of the Jovian clouds would appear to do. But this is far from being all. This great rift *grew* lengthening out, until it stretched across the whole face of the planet, and it grew in a very strange way; for its two ends remained at unchanged distances from the planet's equator, *but the one nearest to the equator travelled forwards* (speaking with reference to the way in which the planet turns on its axis), the rift thus approaching more and more nearly to an east and west direction.' 'As pictured on April 9, the dark rift cannot be estimated at less than a hundred thousand

[1] *Familiar Science Studies*, p. 68.

miles in length, or long enough to extend four times round the Earth's equator.'[1]

It will thus be seen that the cloud covering of Jupiter travels round its axis at different rates of speed—faster at the equator, slower as it nears the poles. The fact has, as we shall afterwards see, a most important bearing on the nature of the planet.

Mr. Proctor observes upon the very obvious circumstance that these belts of Jupiter indicate a condition totally different from the cloud system of Earth:—

'We have been so long accustomed to look upon the belts of Jupiter as due to clouds resembling terrestrial clouds in origin and behaviour, that it may seem surprising to the reader to be told that if the belts really consist of clouds, these must be wholly unlike any with which our meteorologists are acquainted.'[2]

The opinion he had then formed was that the belts indicated

'the existence of a very extensive vapour-laden atmosphere. The dark belts must not be considered as the true cloud belts, because it must be remembered that we look upon the reverse side of the skyscape presented during the day to the Jovials, so that when they see densely compacted dark clouds, we see the light which these clouds have intercepted; and, on the other hand, where they see clear spaces, the light which reaches them is not reflected to us without a considerable loss of brilliancy. Thus the dark belts of Jupiter are those regions where, if at all, we see the true surface of the planet.'[3]

The views thus expressed, however, seem not alto-

[1] *Orbs around us*, p. 131.   See also *Familiar Science Studies*, p. 62.
[2] *Orbs around us*, p. 129.   [3] *Other Worlds*, p. 130.

gether in harmony with those which further reflection suggested to Mr. Proctor.

These cloud layers, which cover the whole surface of Jupiter, are of enormous depth. Mr. Proctor says it would be difficult to express an opinion as to the actual depth, but would assign a minimum of 6000 miles, although strongly of opinion that in reality it is still greater.[1] The mere fact that an atmosphere is recognisable from our distant station is remarkable. 'It is certain that, except by the effects produced when clouds form and dissipate, our terrestrial atmosphere could not be recognised at Jupiter's distance with any telescopic power yet applied.'[2] The atmospheric cloud, or all that we see of Jupiter, consists 'of semi-transparent materials,'[3] and the darker bands may be due to the cloud layers which lie beneath, but 'neither the dark belts nor the bright ones are opaque.'[4]

The spectroscope has afforded some information regarding the composition of this atmosphere. Roscoe says:—

'In the spectrum of Jupiter lines are seen which indicate the existence of an absorptive atmosphere about this planet. These lines plainly appeared when viewed simultaneously with the spectrum of the sky, which at the time of observation reflected the light of the setting sun. One strong band corresponds with some terrestrial atmospheric lines, and probably indicates the presence of vapours similar to those which float about the Earth. Another band has no counterpart amongst the lines of

[1] *Familiar Science Studies*, pp. 71, 76.
[2] *Our Place among the Infinities*, p. 110.
[3] *Familiar Science Studies*, p. 68.   [4] *Ibid.* p. 69.

absorption of our atmosphere, and tells us of some gas or vapour which does not exist in the Earth's atmosphere.'[1]

Schellen's account is a little more ample, but to similar effect:—

'As early as 1864, Huggins discovered some dark lines in the red portion of Jupiter's spectrum which were not coincident with any of the Fraunhofer lines of the solar spectrum, and among them is one that does not occur among the telluric lines.'[2]

(In a footnote it is mentioned that Mr. Le Sueur in 1869 saw the absorptive lines as they are described by Huggins.) Schellen adds:—

'Browning distinctly recognised these lines early in 1870, and thinks that in the green part of the spectrum near the yellow several fine dark lines occur, which are coincident with those occasioned by the vapours of the Earth's atmosphere, and which are generally visible in the corresponding portion of the solar spectrum when the Sun is near the horizon. If it be supposed that Jupiter is in any way self-luminous, these lines may be occasioned by such elements in the planet as are not to be found in the Sun, or if present in the Sun, have not been revealed to us by any effect of absorption.'

And dealing with Saturn, Schellen says:—

'The observations of Janssen, which have been supported by Secchi, have since shown that aqueous vapour is probably present both in Jupiter and Saturn.'[3]

Thus while 'the telescope when applied to Jupiter shows us nothing which can be compared with any known features of our own Earth,'[4] we find from the spectroscope that the atmosphere with which it is surrounded is partly akin to that of the Earth, and it

---

[1] Roscoe's *Lectures*, p. 230.   [2] Schellen, *Spectrum Analysis*, p. 482.
[3] *Ibid.* p. 483.   [4] *Orbs around us*, p. 126.

contains aqueous vapour; but it contains also unknown elements which do not occur in our atmosphere, whether injurious to life or not cannot be said, nor in the view to be expressed does it signify.

Now arises the question to what must we attribute this peculiar atmosphere with its attendant phenomena.

And, in the first place, it is not due to or produced by the present action of the Sun. At the immense distance which separates the two great bodies of our system, the light and heat received by the planet from the Sun is only at most 1/25th of that received by the Earth. It is obvious that any influence arising out of the Sun's heat and light must therefore be weak and imperceptible.

'Let it be remembered,' says Mr. Proctor, 'that supposing Jupiter's globe even to be wholly covered with water, yet a sun twenty-five times farther off than ours could not by any possibility load his atmosphere with the enormous masses of vapour actually present in it. Let it be remembered, further, that the relatively sluggish action of the Sun upon Jupiter could not by any possibility give rise to atmospheric disturbances so tremendous as those which are evidenced by the rapid changes of figure of his cloud bands. When to this we add the relative minuteness of the seasonal changes on Jupiter, we see at once that unless some other cause than solar action were at work, the condition of Jupiter's atmosphere ought to be very much calmer than that of the Earth's.'[1]

'The seasonal changes on Jupiter correspond to no greater *relative* change than occurs in our daily supply of solar heat from about eight days before to about eight days after the spring or autumn equinox. It is incredible that so slight an effect as this should produce

[1] *Other Worlds than Ours*, p. 139.

those amazing changes in the condition of the Jovian atmosphere, which have unquestionably been indicated by the varying aspect of the equatorial zone. It is manifest that, on the one hand, the seasonal changes should be slow and slight so far as they depend on the Sun; and that, on the other, the Sun does not rule so absolutely over the Jovian atmosphere as to cause any particular atmospheric condition to prevail unchanged for years.' [1]

Mr. Proctor recurs to the subject in dealing with the cloud rifts. Apart from the evidence afforded

'of the swift motions of the cloud masses enwrapping Jupiter from a velocity of 151 per hour, exceeding that of the most tremendous hurricanes upon Earth, it has always seemed to me,' he says, 'that this one series of observations should suffice of itself to show that the phenomena of Jupiter's cloud-laden atmosphere are not due to solar action. For the rift itself continued, and the changes affecting it continued whether Jovian day was in progress or Jovian night. For one hundred Jovian days or more, and for one hundred Jovian nights, the great cloud masses on either side of the rift remained in position opposite each other, slowly wheeling, but still continuing face to face as their equatorial ends rushed onwards at a rate fourfold that of a swift train, even measuring their velocity only by reference to the ends remote from the equator, and regarding these as fixed. Probably the cloud masses were moving still more swiftly with respect to the surface of the planet below.' [2]

The belts formerly were associated in our minds with trade winds, and apparently were supposed to spring from similar causes.[3] But Mr. Proctor [4] shows

---

[1] *Our Place among Infinities*, p. 122. [2] *Familiar Science Studies*, p. 63.
[3] Laplace (*System of World*, i. p. 80) says: 'They appear to be clouds which the winds transport with various velocities in an extremely agitated atmosphere.'
[4] *Orbs around us*, p. 133, also p. 142.

that Jupiter's clouds and the movements in them are not due to winds. The motion of the great rift already referred to was 'the direct reverse' of the motion of air currents producing the trade winds, while the rate of motion in Jupiter vastly 'exceeded anything we recognise in the trades or counter trades;' a velocity far exceeding that of such a hurricane upon Earth as would here produce universal desolation, and approaching 'those solar cyclones of which the spectroscope has given us such startling intelligence.'[1]

And with reference to the astounding rapidity of dispersion of the spot observed by South, already mentioned, Mr. Proctor says that to suppose winds had carried cloud masses athwart the opening would imply a rush at the incredible rate of 20,000 or of 40,000 miles per hour, which would exceed more than a hundred times (taking the least number) the velocity of our most tremendous hurricanes.[2]

But it is in the appearances indicating the real condition of Jupiter we must look for explanation.

Mr. Proctor has, in a chapter of *Our Place among Infinities*,[3] very satisfactorily shown what that must be. Starting with the fact that the atmosphere of Jupiter is necessarily attracted by the mass of the planet, and that owing to its greater mass the power of gravitation is more than twice what it is upon Earth, he calculates that, assuming the cloud layers are at least 100 miles in depth, the pressure would exceed that at our sea level so many times that the numbers representing the proportion contain twenty-one figures. This would be infinitely beyond the limit at which atmospheric pressure cannot be increased without changing the compressed

[1] *Orbs*, p. 134.   [2] *Flowers of the Sky*, p. 205.   [3] P. 112 et seq.

air into liquid form, and he enters into illustrations to show that the pressure would produce a density inconsistent with its actual density were the atmosphere of Jupiter existing at a temperature not greatly differing from that of our own. The only way by which the difficulty can be avoided is by assuming an exceedingly high temperature. Assume this, and 'we no longer find any circumstances which are self-contradictory or incredible.'[1] A parallel is found in the case of the Sun, where the enormous pressure is counterbalanced by the enormous heat, whence ' we may safely adopt the conclusion that *Jupiter is intensely heated, though not nearly to the same degree as the Sun.*'

Mr. Proctor—observing that

' many lines of evidence, and some of them absolutely demonstrative, in my opinion point to the conclusion that Jupiter is an orb *instinct with fiery energy, aglow, it may well be, with an intense light,* which is only prevented from manifesting itself by the cloudy envelope enshrouding the planet '—finds in this hypothesis 'the means of readily interpreting what otherwise would appear most perplexing.'[2]

Thus he is inclined to regard the ruddy glow of Jupiter's equatorial zone, during the period of disturbance already mentioned,[3] as ' *due to the inherent light of glowing matter* underneath his deep and cloud-laden atmosphere.'[4]

In the same paper[5] he points to the circumstance that the fourth and outermost satellite of Jupiter in crossing his disc looks perfectly black,—though doubtless possessed of reflective power equal to that of our own Moon,—as evidencing the greater luminosity of Jupiter,

---

[1] Proctor's *Infinities*, p. 119.   [2] P. 121.   [3] P. 123.
[4] See also Newcomb, p. 334.   [5] P. 126.

and consequently that Jupiter sheds a light stronger than what would arise from reflection merely. Elsewhere Mr. Proctor has referred to the estimates of Zöllner, the eminent photometrician, as showing that he sends

'much more light to us than a planet of equal size, and constituted like Mars, the Moon, or the Earth, could possibly reflect to us if placed where Jupiter is. Whereas Mars reflects but one-fourth of the light he receives, Jupiter reflects more than three-fifths. The Moon sends less than a fifth; Saturn, Jupiter's brother giant, more than a half.'[1]

And with this the calculations of others agree.

'Indeed, if Jupiter do not shine in part by native light, his surface must possess reflective powers nearly equal to those of white paper,' which cannot be, as a large proportion of his surface is of a dull coppery hue. 'It follows as at least highly probable that Jupiter shines in part by his own light; and this being admitted, we cannot but regard it as highly probable that *the mass of the planet must be intensely hot.*'[2]

Another concurring fact is, that whereas were Jupiter's atmosphere of the same temperature as our Earth's it would be excessively compressed, 'the *observed mobility* of Jupiter's cloud envelope and other circumstances indicate that this enormous compression does not exist.'

Indeed, the most striking proofs of Jupiter being in a condition analogous to that of the Sun lie in its resemblance to the Sun. In density it nearly coincides, the difference being only fractional. In the formation and closing up of spots it is not unlike. In the more rapid movement of the atmosphere or cloud surface at the equator than at parts nearer the poles, it evidences

[1] *Other Worlds than Ours*, p. 143. [2] *Ibid.* p. 144.

a most striking likeness in condition to the Sun. Indeed it admits of no doubt that Jupiter is in a sense a minor sun, although his clouds are not, from defect of sufficient heat in the planet, in themselves luminous, or they are only faintly so.

Nor is this surprising when we consider the origin of the planetary system as disclosed by the Nebular Hypothesis, which demonstrates that 'every planet must once in its life pass through the fiery stage of planetary existence,'[1] and it is obvious that a planet so much larger than our Earth must require vastly longer time to cool down from its nebulous heat. Mr. Proctor calculates that if Bischoff be right in assigning 340,000,000 years to the era of the Earth's past since it was a molten globe, it would require that Jupiter should take about seven times as long, or about 2,380,000,000 years to arrive at the present condition of the Earth; or supposing Bischoff's calculation to be overrated, and taking it at a tenth, it would leave 238,000,000 years—in either case it would be '*behind the Earth as respects this stage of development.*'[2] But this era was preceded by others which are usually considered to have lasted much longer.

'The Earth was once a mighty ring surrounding the Sun, and had to contract into globe form, a process requiring many millions of years. When first formed into a globe she was vapourous, and had to contract— forming the Moon in so doing—until she became a mass, first of liquid, then of plastic half-solid matter, glowing with fire, and covered with tracts of fluent heat. Here was another stage of her past existence, requiring probably many hundreds of millions of years. Jupiter and Saturn had to pass through similar stages of development, and required many times as many years for each

[1] *Flowers of the Sky*, p. 195.   [2] P. 196.

of them. Is it then reasonable to suppose that they have arrived at the same stage of development as our Earth, or indeed as each other?'[1]

This being so, we cannot be surprised to find that the cloud masses of Jupiter are subject to great disturbance, and sometimes the disturbance has been in apparent sympathy with disturbance in the Sun;[2] or, as it may be put, 'whenever Jupiter is so situated as to be at work most effectively in disturbing the Sun, then he is himself most disturbed.'[3]

Mr. Proctor[4] has accounted for the formation of the belts of Jupiter by violent up-rushes of vapour from vast depths below his visible surface; and similarly he accounts for the white, egg-shaped clouds which have been mentioned 'by supposing them due to explosive action casting up enormous masses of vapour into the higher regions of Jupiter's atmosphere.' He has entered most fully into the question in the chapter on Jupiter in his *Familiar Science Studies*,[5] in the course of which he attributes the belts to vertical up-rushes and down-rushes; which indeed would, to a certain extent, offer another analogy to the condition of the Sun, according to Mr. Lockyer's theory, although the analogy does not in all respects strictly hold. Let me add, that as the disturbances seem to be principally at the equator of Jupiter, it may be that at one time there surrounded the planet a ring or rings like those of Saturn, which, having collapsed and fallen into the planet, excited it to more violent action in that quarter, and produced in the equatorial belt that pearly whiteness by which it is distinguished beyond the others.

[1] *Flowers of the Sky*, p. 197. [2] *Orbs around us*, p. 137. [3] P. 141.
[4] *Orbs around us*, p. 143. [5] *Familiar Science Studies*, p. 65 et seq.

It is, however, not necessary to discuss these matters. It is sufficient that we are 'led to the conclusion that Jupiter is still a glowing mass,—fluid probably throughout, *still bubbling and seething with the intensity of the primeval fires*,—sending up continually enormous masses of cloud to be gathered into bands under the influence of the swift rotation of the giant planet.'[1]

The foregoing observations have been rested on Mr. Proctor's views and statements. Mr. Matthieu Williams, who published his *Fuel of the Sun* in 1870, or about simultaneously with Mr. Proctor's *Other Worlds than Ours*, expressed (and it may therefore be supposed independently) opinions regarding Jupiter somewhat akin, though not in every respect coinciding.

'I have little doubt,' he says,[2] as the result of his speculations, 'that Jupiter is still red hot, or rather white hot; that a vast depth of aqueous and other vapour surrounds it; and that these, together with the free oxygen and free nitrogen, form a very much greater atmosphere than that which I have calculated. I think it extremely probable that the temperature of dissociation of water has been reached by the original atmospheric compression of Jupiter; that he must have manifested some degree of general solar phenomena; and that, if we could see him shaded from the solar rays, he would appear like a phosphorescent, or rather a fluorescent ball, by the illumination of his vapourous envelope, due to the light it absorbs from the glowing world within.'

Such being the fiery condition of Jupiter, we are led necessarily to the further conclusion (and this is all we are concerned with here), that, using the words in which Mr. Proctor expresses his own opinion, 'we must of

[1] Proctor's *Other Worlds*, p. 140.   [2] *Fuel of Sun*, § 286, p. 189.

course *dismiss the idea that the giant planet is at present a fit abode for living creatures.*'

Formerly the moons of Jupiter were looked upon as affording 'perpetual moonlight' to his cartilaginous inhabitants; but Jupiter being found in no condition to accommodate inhabitants of any kind, it was perhaps not unnatural for those wedded to plurality to transfer these inhabitants to his satellites, and find room and comfortable quarters for them there.

'If Jupiter be still in a sense a sun,—not, indeed, resplendent like the great centre of the planetary scheme, but still a source of heat,—is there not excellent reason for believing that the system which circles around him consists of four worlds where life, even such forms of life as we are familiar with, may still exist? Those four orbs, which our telescopes reveal to us as tiny points of light, are in reality globes which may be compared with the four worlds that circle nearest the Sun.'[1]

Mr. Proctor repeats this view in a chapter of *The Expanse of Heaven*, p. 85.

It appears sufficient answer to say that 'these moons seem to be made of even lighter material than the planet itself, for the densest would be much more than outweighed by half its bulk of water.'[2]

Another alternative has been to suppose that at some future day Jupiter may be fitted for life. 'The processes we see at work out yonder may be fitting him for the support of myriads of races of living creatures. For anything we know to the contrary, he may be passing through stages which our own Earth has long since passed through.'[3] But whether before our Sun has

[1] *Other Worlds*, p. 141.   [2] *Orbs around us*, p. 126.   [3] P. 145.

cooled down and become a black orb, sending out neither heat nor light to any of his planets, Jupiter shall have cooled down to the condition of the Earth, we cannot tell,—and Mr. Matthieu Williams is of opinion that he has probably reached his permanent temperature,—but this we know (and it may perhaps be considered very cruel and hard-hearted to crush out even this hope), that the facts which have been stated, and more especially his great distance from the Sun, and the inconceivable cold in which he would be wrapped, are quite sufficient to destroy the thought that Jupiter in that condition could become habitable.

SATURN.

Dr. Dick,[1] in conformity with his plurality views, thus observes upon Saturn:—

'This globe, which appears only like a dim speck on our nocturnal sky, may be considered as equal to six thousand worlds like ours; and since such a noble apparatus of rings and moons is provided for the accommodation and contemplation of intelligent beings, we cannot doubt that it is replenished with ten thousand times ten thousand of sensitive and rational inhabitants, and that the scenes and transactions connected with that distant world may far surpass in grandeur whatever has occurred on the theatre of our globe.'

Little help to the present question is indeed to be found in the notions which, till ten or twelve years ago, prevailed regarding these outer planets. Even Mr. Proctor, who published his book on *Saturn* in 1865, or four years before the ideas occurred to him which, with reference to Jupiter, have been explained, adopted in that work those older conceptions, although his views regarding Saturn subsequently experienced the same change. But before stating those his older opinions, it will be well to note one or two facts relative to the planet now to be considered.

Saturn is nearly twice as distant from the Sun as Jupiter is, or about ten times as far off as the Earth is,

[1] *Celestial Scenery*, p. 166.

and consequently the light and heat it receives from the Sun are greatly less; for while the heat and light received by Jupiter are about 1/25th, the heat and light received by Saturn are only 1/90th of that received by the Earth. The eccentricity of its orbit is more than twice that of the orbit of Jupiter, the difference between its aphelion and its perihelion being nearly 100 millions of miles; so that the difference between the light and heat at the one position is very greatly less than it is at the other. Its year extends to $29\tfrac{1}{2}$ of our years, but its rotation on its axis approximates in rate to that of Jupiter, its day being only half an hour longer. The inclination of its axis does not largely differ from that of the Earth; so that there would, if climate permitted, be a possibility of the four seasons, but they would be each thirty times as long as the seasons of the Earth, and they would be subject to such disturbance by the eclipsing caused by the rings as practically to be neutralized or rendered unserviceable.

The equatorial diameter of Saturn is about 75,000 miles, its polar diameter about 7000 miles less, so that the planet is greatly flattened at the poles. In volume, while Jupiter is 1300 times larger than the Earth, Saturn is only 700 times larger. But the density of Saturn is greatly less than that of Jupiter; for while the mass of Jupiter is equal to nearly 340 times that of the Earth, the mass of Saturn is only above 90 times. Its specific density is about 1/8th (Lardner has it 1/10th) that of the Earth. It is indeed the lightest of all the planets; and one result is, that notwithstanding its vastly greater bulk, gravity upon Saturn is nearly equal to gravity upon the Earth, although gravity is said to be 1/5th greater at his poles than at his equator.

SURFACE.

Saturn resembles Jupiter in being marked with belts; but it differs from its big brother and from all the other planets in the great rings which encircle it. Besides these rings, it is attended by seven satellites. I shall afterwards have to refer more particularly to these circumstances.

These are all facts which have long been well known, and in the light of them Mr. Proctor thus wrote in 1865:—

'It has been considered probable that the appearance of Saturn's surface differs greatly from that of our Earth. And this for two reasons: In the first place, it is urged that his density being so small, he must be composed of materials very much lighter than, and therefore very different from, those composing our Earth; and, in the second place, that fluids upon his surface must either be of less density than the planet, and therefore very different from our oceans, or if of greater density, must all be collected in one hemisphere. Saturn's globe, however, may be hollow, and the mean density of the materials of this hollow globe not very different from the mean density of the materials composing our Earth. And again, it has not been established by rigid mathematical inquiry that oceans upon a planet of Saturn's figure would necessarily be collected wholly, or almost wholly, in one hemisphere if their density exceeded that of the planet. On the contrary, it appears probable that fluid masses on the surface of such a planet would tend to form two vast polar oceans, since gravity is so much greater at Saturn's poles than at his equator. But even if it were proved that the former arrangement must inevitably subsist, what, after all, is such an arrangement but an almost exact counterpart of what is observed in our own Earth. It is true that tidal waves could not sweep round such an ocean as round the oceans that surround the Earth; but an ocean whose tides are are ruled by eight satellites and restrained by the attractions of a stupendous ring may require arrangements altogether different in this respect from those prevailing on our Earth. The appearance of Saturn, however, is not favourable to the supposition that the ocean masses on his surface are confined to a single hemisphere, for the bright bands on Saturn's disc, which are probably vast belts of clouds drawn from oceans upon his surface, are found equally in the northern and southern hemispheres, and extend completely round Saturn's globe.

'The climatic conditions on the surface of Saturn undoubtedly differ in the most striking manner from those which prevail on the Earth.'[1]

Mr. Proctor then proceeds at considerable length to consider them, and thus closes the chapter on the 'Habitability of Saturn,' in which the above remarks occur:—

'The result of the examination of the probable physical conditions and phenomena subsisting in Saturn does not appear to favour the supposition that the planet is a suitable habitation for beings constituted like the inhabitants of our globe. The variation of gravity, the length of the Saturnian year, and the long protracted eclipses caused by the ring, are the circumstances that seem to militate most strongly against such a supposition. Over a zone near the Saturnian equator these circumstances have less effect, however; and it is not impossible that arrangements unknown to us prevail in Saturn which may render other parts of his surface habitable as we should understand the term: "The very combinations which convey to our minds only images of horror may be in reality theatres of the most striking and glorious displays of beneficent contrivance."

'On the general question of the habitability of the system that circles about Saturn we have no means of forming an opinion. From the analogy of our Moon it appears highly probable that no part of the system is habitable by such creatures as inhabit our Earth.'[2]

[1] *Saturn*, by R. A. Proctor, p. 158. [2] *Ibid.* p. 185.

There has been discussion about the duration of the eclipses on the face of Saturn caused by the rings, and the consequent effect upon the planet; but, indeed, it is unnecessary to enter upon this or any kindred question involving the supposition of a physical condition of the planet analogous to that of the Earth, because the views now entertained, and pretty fully stated, in dealing with Jupiter are as applicable to Saturn, manifesting it to be in a like predicament; or in Mr. Proctor's own words, 'that an intense heat must in all probability prevail on the great globe of Saturn,'[1] rendering it entirely unsuitable as an abode for living creatures.

I will just shortly notice the different facts which converge to this conclusion.

1. There is the small density of the planet combined with its magnitude. This cannot be due to solar action, for the planet is still farther removed from the Sun than Jupiter is, and the Sun has, by so much, less power upon it. But we have seen in the case of Jupiter that the enormous pressure is inconsistent with the small density, except upon the footing of assuming an exceedingly high temperature to exist in the body or interior of the planet. The bulk of Saturn is little more than half that of Jupiter, but the same principle applies, and it applies the more seeing the density is so much less, being about the half. Perhaps I might venture to say that some explanation of the smaller density may be found in the counter attraction of the rings, or it may be in the greater heat which the rings serve to enclose, causing a more considerable distension. Whether this be so or not the fact remains, proving a state of intense heat.

[1] *Other Worlds than Ours*, p. 160.

Mr Matthieu Williams[1] holds the temperature assignable to the compression to be 590° Fahr.; and, according to Dalton, it should be 625° Fahr. He says:—

'The specific gravity of Saturn being still lower than that of Jupiter, I am irresistibly led to deny its solidity, and even to go further, and maintain that we are justified in believing that it is still hotter than Jupiter, and that the proportion of its gaseous constituents must also be greater.' And he goes on to say that 'if Saturn and the other light planets are thus mainly composed of gaseous matter, the mechanical work of gravitation exerted in their aggregation and compression must have produced a vastly greater evolution of heat than I have calculated as due to the compression of merely that portion of the general atmospheric medium constituting their supplementary envelope, and this high temperature, after being reduced to a certain point, would be maintained by the gravitation of their satellites.' 'Saturn must be even more completely than Jupiter a miniature or model sun.'

2. It is, like Jupiter, girded with belts.

'In November 1793, Herschel obtained a favourable view of the Saturnian belts with his 40 feet reflector. He observed a broad and brilliant white belt of nearly uniform width, covering the equatorial regions; next to it he observed a broad dark belt of a yellowish colour' (Mr. Proctor, in a footnote, says modern telescopes exhibit the dark belts as of a faint greenish colour), 'divided into three unequal bands by two narrow and somewhat irregular white streaks less brilliant than the equatorial belt.'[2]

In this description there will not fail to be noticed the resemblance to Jupiter in the bright equatorial belt, pointing to a common condition. Indeed the belts

'resemble those of Jupiter in their general shape, and

[1] *Fuel of Sun*, P. 196   [2] Proctor's *Saturn*, p. 60.

also in their colour. The dark belts near the equator are of a faint brown or ruddy tinge, those near the pole bluish or greenish grey, while the bright belts are yellowish—the equatorial belt being the brightest of all, and almost white. The poles are commonly dusky and even sombre in hue.'[1]

The resemblance does not stop there, for it appears 'the belts change in aspect much as those of Jupiter have been observed to do.'

These facts reveal an analogous condition underlying the belts.

3. Saturn evidently emits inherent light over and above the light it reflects.

'The light sent to us from Saturn bears a much greater proportion to the amount of solar light actually received by the planet than is observed in the case of Mars or the Moon, and so nearly approaches the proportion noticed in the case of Jupiter as to lead to the same inference.'[2]

As already stated, according to Zöllner, while the Moon reflects less than a fifth of the light she receives, Saturn reflects more than a half, or more than double the Moon's reflection.

4. There are spots on Saturn, and it was by watching them that Herschel[3] found the period of its revolution on its axis. But probably owing to the great distance and consequent difficulty of observation there is nowhere, so far as I have seen, any minute description of their appearance and changes.

5. But a fact quite as cogent, if not more so, is that Saturn occasionally appears distorted. This was observed by Sir William Herschel for the first time in

[1] *Other Worlds*, p. 155. [2] *Ibid.* [3] *Saturn*, p. 57.

1805. For[1] its shape suffers alteration, the parts midway between the pole and the equator of the planet bulging out and becoming 'square-shouldered.' This remarkable appearance has since been noticed by several astronomers, and it practically amounts to this, that its figure is susceptible of terrible variation, so that we must 'accept the astounding conclusion that the giant bulk of Saturn is subject of throes of so tremendous a nature as to upheave whole zones of his surface five or six hundred miles above their ordinary level.'[2] Somewhat of the same appearance has been noticed in Jupiter, but apparently not so well marked. It indicates the forces of heat moving within with appalling power.

6. *Spectrum Analysis* has shown that there is a resemblance between Jupiter and Saturn.

'The comparatively faint spectrum of Saturn has been examined by Huggins, who observed in it some of the lines characteristic of Jupiter's spectrum. These lines are less clearly seen in the light of the ring than in that of the ball, whence it may be concluded that the light from the ring suffers less absorption than does the light from the planet itself. The observations of Janssen, which have been supported by Secchi, have since shown that aqueous vapour is probably present both in Jupiter and Saturn. Secchi has further discovered some lines in the spectrum of Saturn *which are not coincident with any of the telluric lines,* nor with any of the lines of the solar spectrum produced by the aqueous vapour of the Earth's atmosphere. It is not improbable, therefore, that the *atmosphere of Saturn may contain gases or vapours* which do not exist in that of our Earth.'[3]

All these circumstances combine to show that Saturn is, like Jupiter, a planet differing from this Earth, and

---

[1] *Flowers of the Sky,* p. 222.   [2] *Other Worlds,* p. 157.
[3] Schellen, p. 483.

is still burning with nebulous heat, altogether inconsistent with the idea of being habitable.

This being so, the resort is to suppose it to be a sun to the eight satellites which surround it.

Mr. Proctor thus expresses his views on this subject:—

'It seems to me that, apart from the reasoning already adduced, we have to choose between two views of the Saturnian system. Either the scheme of satellites and the system of rings are intended to subserve some useful purpose with respect to Saturn, or Saturn subserves some useful purpose with respect to these systems. Now the satellites can supply very little light to Saturn. All together (if they *could* be all full together) they would supply but a sixteenth part of the light which we receive from our Moon when she is full. How so insignificant a supply of reflected light can make up to Saturnians for the fact that the direct supply of solar heat is but one ninetieth of that which we receive, I leave the believers in Saturn's habitability to explain. But the ring system, which also has been spoken of as supplementing the deficiency of solar light, does just the reverse. It deprives the Saturnians for long periods together, in some regions for several successive years, of the light they would otherwise receive. And this it does in the winter of those places. At this time also it reflects no light to them during the night. In summer the rings do not cut off any of the Sun's light, and they shine at night with a considerable degree of brightness, marred only by the circumstance that at midnight the great shadow of the planet falls on nearly the whole of the visible part of the ring. But no supply of reflected light during the summer nights can compensate for the deprivation of the whole of the Sun's direct light *in winter* for several of *our* years together.

'We seem compelled, then, to adopt the view that Saturn subserves useful purposes to the worlds which circle round him. To these he certainly supplies much reflected light, and possibly a considerable proportion of inherent light. He probably warms them in a much greater degree. And it seems no unworthy thought respecting him, that even as he sways them by his attractive energy, so he nourishes them as a subordinate Sun by the heat with which his great mass is instinct. If our Sun, so far surpassing all his dependent worlds in mass, yet acts as their servant in such respects, we may reasonably believe that Saturn and Jupiter act a similar part towards the orbs which circle round them.[1]

This theory, it will be seen, is built upon the idea that the planets and other heavenly bodies must necessarily subserve a useful purpose, or rather such a useful purpose as we can conceive; a view which has been previously disposed of in dealing with the general question.

With regard to the satellites themselves, we know too little to say much about them. They would seem to resemble our Moon in presenting the same face to the planet throughout their revolutions (and indeed this is an observation which is thought to apply to the satellites of all the planets), and they may resemble the Moon, therefore, in condition. But if they possess an atmosphere, it may resemble that of their parent orb, which, according to *Spectrum Analysis*, differs from the life-sustaining atmosphere of Earth, which would of itself, apart from all other circumstances, render them uninhabitable; a conclusion, however, which is sufficiently forced upon us by the fact of the enormous distance separating them from the Sun depriving them of the heat necessary for life—a reason which would be equally applicable to the moons of all the exterior planets.

[1] *Expanse of Heaven*, p. 102.

I have not considered the case of the rings, because they are apparently either a vast heap of small meteoric bodies, or if not so, are probably composed of gases. In either case they cannot be deemed habitable. The widening of the rings seems to indicate a possibility of their descending to the planet, and thus maintaining its heat and agitation indefinitely.

URANUS.

TILL a century ago, Saturn had been regarded as the outermost planet of the solar system. It was little thought that far beyond it another was ploughing its lonely way in the darkness and depths of space, freighted, according to Dr. Dick, with a constant cargo of 1,077,568,000,000, or more than a billion of souls. Perhaps the reader, after all that has been advanced, may be inclined to discount a little from this precise estimate; but it will be my business to help him to decision.

On 13th March 1781 this planet was discovered by Sir William Herschel, and at once the industrious ardent amateur leapt from obscurity to fame. The discovery, his biographer says, 'was the most striking since the invention of the telescope. It had absolutely no parallel, for every other major planet had been known from time immemorial.'[1] Herschel named it in honour of George III. the *Georgium Sidus*. Others honoured it more appropriately with his own name. Ultimately it was called Uranus, the most ancient of the gods, whose son Saturn, and grandson Jupiter, had already obtained a lasting establishment in the starry firmament.

This planet rolls in its distant course more than 1800 millions of miles from the Sun, being nearly

[1] See the all too scanty *Life of Herschel*, by Holden, p. 48 et seq.

twenty times as far removed from it as is the Earth. In size it is not so large as Saturn; its diameter of about 35,000 miles being nearly half that of the other, the polar being 1/9th less than the equatorial diameter, but in density it is slightly greater, or as 167 is to 131. In mass it is nearly fifteen times greater than the Earth. Its gravity is about 3/4ths that of the Earth, or ·76—the Earth being = 1. The light and heat received by Uranus from the Sun are 1/390th of that received by the Earth, and therefore must be faint indeed. Did it resemble the Earth, it has been calculated that the cold at its surface would be 122° Fahr. below freezing.[1] Its year consists of eighty-four of our years, but it has not been possible to observe the length of its day. It has been thought by Sir William Herschel and other astronomers, but not by all, that it is flattened at the poles, Mädler calculating the flattening to amount to 1/10th, which, if correct, would be suggestive of a rapid rotation.[2] It is attended by six moons.

It will at once be apparent that if Jupiter and Saturn be uninhabitable in respect of distance from the Sun, so *multo majus* must Uranus be.

There is a striking peculiarity in the inclination of the plane of Uranus' equator to the path in which he travels, upon which Mr. Proctor in *More Worlds than One*[3] dwells to show the effect upon its seasons or climatic condition, rendering life upon it all the more impossible. But in the face of the more serious objection that it is *in pari casu* with Jupiter and Saturn as

---

[1] Dick's *Celestial Scenery*, p. 194.
[2] Guillemin's *Heavens*, p. 228.
[3] P. 167.

regards its fiery constitution, it is superfluous to enter upon matters of this kind.

The same reason which has been urged in the cases of Jupiter and Saturn applies to Uranus, that the amount of its density, combined with its bulk, prove it to be still in a state of nebulous heat.

'The density of this planet,' says Mr. Williams,[1] 'is intermediate between that of Jupiter and Saturn, and if my views of the constitution of these two planets are sound, Uranus should, like them, be mainly composed of gaseous matter, and must have the general constitution of a minor sun. Much of the heat evolved by the compression of its gaseous constituents (I refer now to planetary, not to atmospheric gases) must be still maintained, and the source of this high permanent temperature ought to be traceable to the amount of disturbing reaction of its satellites.'

Uranus is far too distant to admit of observation of its surface. Whether it has belts, or spots, or streaks, however probable, seems to be unknown. But it has been examined by the spectroscope, and the results of that examination are most important.

'The spectrum of Uranus,' says Schellen,[2] 'which has been investigated by Secchi, appears to be of a very remarkable character. It consists mainly of two broad black bands—one, m, the greenish-blue, but not coincident with the F line; and the other, n, in the green, near the line E. A little beyond the band n, the spectrum disappears altogether, and shows a blank space, q, p, extending entirely over the yellow to the red, where there is again a faint reappearance of light. The spectrum is therefore such a one as would be produced were all the yellow rays extinguished from the light of the Sun. The dark sodium line, D, occurs, as is well known, in the

[1] *Fuel of Sun*, § 315. [2] Schellen, p. 484.

part of the spectrum occupied by this broad non-luminous space. Is this extraordinary phenomenon, therefore, to be ascribed to the influence of this metal? or is the planet Uranus, which has a spectrum differing so greatly from that of the Sun, self-luminous? Has the planet not yet attained that degree of consistency possessed by the nearer planets, which shine only by the Sun's light, and, as the photometric observations of Zöllner lead us to suppose is possible, is still in that process of condensation and subsequent development through which the Earth has already passed? These are questions to which at present we can furnish no reply, and the problem can only be solved by additional observations of the strange characteristics exhibited by this spectrum.'

Dr. Huggins,[1] by means of a more powerful telescope, subsequently examined the planet, and found the spectrum to be complete, no part being wanting, so far as the feebleness of its light permitted it to be traced.

'But,' observes Mr. Proctor,[2] 'there are six dark bands or strong lines indicating the absorptive action of the planet's atmosphere. One of these strong lines corresponds in position with one of the lines of hydrogen. Now it may seem at a first view that since the light of Uranus is reflected solar light, we might expect to find in the spectrum of Uranus the solar lines of hydrogen. But the line in question is too strong to be regarded as merely representing the corresponding line in the solar spectrum. Indeed, Dr. Huggins distinctly mentions that the bands produced by planetary absorption are broad and strong in comparison with the solar lines. We must conclude, therefore, that there exists in the atmosphere of Uranus the gas hydrogen, sufficiently familiar to us as an element which appears in combination with others, but which we by no means recognise as a suitable constituent (at least to any great extent) of an atmosphere, which

---

[1] Schellen, p. 485.
[2] *Light Science for Leisure Hours*, 2d series, p. 144.

living creatures are to breathe. And not only must hydrogen be present in the atmosphere of Uranus, but in such enormous quantities as to be one of the chief atmospheric constituents.'

After adverting to the apparent absence of carbonic acid, and to the small amount, if any, of aqueous vapour, Mr. Proctor proceeds:—

'We certainly have a strange discovery to deal with. If it be remembered that oxygen, the main supporter of such life as we are familiar with, cannot be mixed with hydrogen without the certainty that the first spark will cause an explosion (in which the whole or one or other of the gases will combine with a due portion of the other to produce water), it is difficult to resist the conclusion that oxygen must be absent from the atmosphere of Uranus. If hydrogen could be added in such quantities to our atmosphere as to be recognisable from a distant planet by spectroscopic analysis, then no terrestrial fires could be lighted, for a spark would produce a catastrophe in which all living things upon the Earth, if not the solid earth itself, would be destroyed. A single flash of lightning would be competent to leave the Earth but a huge cinder, even if its whole frame were not rent into a million fragments by the explosion which would ensue.'[1]

It occurs to me that the presence of such overpowering quantities of hydrogen indicate of themselves a condition in Uranus very closely analogous to the Sun, whose atmosphere consists of hundreds of thousands of miles of hydrogen, in one portion incandescent, in another cool.

Mr. Proctor adds:—

'Under what strange conditions, then, must life exist in Uranus, if there be indeed life upon that distant orb. Either our life-sustaining element, oxygen, is wanting, or if

[1] *Light Science,* p. 146.

it exists in sufficient quantities (according to our notions) for the support of life, then there can be no fire, natural or artificial, in that giant planet. It seems more reasonable to conclude that, as had been suspected for other reasons, *the planet is not at present in a condition which renders it a suitable abode for living creatures.*'[1]

Nor can it be a bit more possible in the future, that is, if cooled down to solidity.

With regard to the moons of Uranus little is known beyond a peculiarity in their motion round the planet, explained by Mr. Proctor,[2] which causes them to differ considerably in orbit from that of all the other planetary satellites, and a variation of brightness which has been thought to arise either from the existence of dark tracts on their surface, or from spots or rents in their atmosphere.[3] Possibly they are deadly cold; and if they have an atmosphere, it may partake of that of their death-dealing primary.

[1] *Light Science*, p. 147.   [2] *Ibid.* p. 140.
[3] Grant's *History*, p. 284.

STILL another outside wanderer remains. The extraordinary perturbations of Uranus awoke astronomers to the thought that these must be attributable to the influence of another and unknown planet, and sixty-five years after the discovery of Uranus, Neptune was found. It was the result of high scientific calculation, and the honour of the discovery was divided between Adams, a young English mathematician, and Leverrier, the French astronomer. Adams first in time; Leverrier first in publication. The subject is fully discussed in Professor Nichol's work, *The Planet Neptune: an Exposition and History*. His verdict is, the two stand forth independent—they are PEERS.[1]

With this question of disputed honour we are not concerned here. We have only to dispose of the habitability of the planet, and after what has been said regarding the others, it falls to be dismissed with still less notice.

Neptune is somewhat larger than Uranus, being (according to Mr. Proctor, other writers estimating differently) 37,250 miles in diameter. Its volume is 80 times (more or less) that of the Earth, and its mass 16 or 18 times. Its density is about one-third of that of the Earth, or not quite double that of water, and it

[1] See also Proctor's *Expanse of Heaven*, p. 120, where the matter is briefly mentioned.

is also nearly double that of Uranus, so that perhaps it may be condensing. Its distance from the Sun is 2,800,000,000 miles, or 30 times that of the Earth, and the light and heat it receives are the 1/900th part of that received by the Earth. The inclination of its axis to the orbit is great. It takes 164 years 266 days to perform its long circuit round the Sun. The length of its day cannot be discovered, because it is impossible to note and watch the features of its surface.

In these circumstances there could be no room left for doubt that Neptune cannot possibly be habitable. Mr. Williams[1] observes regarding it that 'considering the pressure of the planetary mass upon itself (eighteen times that of the Earth) the superficial portion of its planetary matter must still be gaseous, and its solid nucleus far below.'

But it is sufficient for present purposes to refer to the revelations of the spectroscope regarding it or its atmosphere. It is thus stated by Schellen:—

'The spectrum of Neptune, which has also been examined by Secchi, *bears a great resemblance to that of Uranus.* It is characterized by three principal bands. The first, which is the faintest, is situated between the green and the yellow, nearly in the centre between D and b; it is of considerable breadth, but very ill defined at the edges. Between this and the red there is a tolerably bright band, with which the spectrum seems suddenly to terminate, and the red is entirely wanting. Secchi is of opinion that the absence of the red is not occasioned by the faintness of this planet, for other stars no brighter than Neptune show the red clearly in the spectrum. The absence of this colour in the spectrum of Neptune must therefore be ascribed to absorption.

'The second absorption band occurs at the line b; it

---

[1] *Fuel of Sun*, p. 209.

is tolerably well defined at the edges, but much fainter
and more difficult of observation than the first band.
The third band is in the blue, and is even fainter than
the second.

'This spectrum is in agreement with the colour of the
planet, which resembles the beautiful tint of the sea. A
peculiar interest attaches to this spectrum from the
coincidence of the dark bands with the bright bands of
certain comets, and with the dark bands of stars of the
fourth type. These bands may possibly be due to carbon;
but accurate measurements are exceedingly difficult, and
can only be attempted on the finest evenings, and with
the use of the most powerful instruments.'[1]

The statement of coincidence with the bands of
comets, and that they may possibly be due to carbon,
may be important, but the most material observation
arising from this analysis is in showing that the
spectrum of Neptune bears a great resemblance to that
of Uranus, so that we may fairly assume that hydrogen
predominates in Neptune also; and therefore, in addition
to being in a gaseous hot state, it possesses an atmosphere in which life is impossible.

Of its two moons little or nothing is known. An
additional one has within the last few months been discovered by Lassell.[2] But both in their case and in that of
the satellites of Uranus, it may be observed that the light
which is shed upon them must be far too feeble to induce
vegetation, even were there heat enough thrown upon
them from Uranus, which there is nothing to warrant in
supposing. Doubtless, if solid, these frigid globes, shiver-
ing with inconceivable cold, wander about the planet by
which they are swayed, which gladly binds them to herself as it pursues its lonely way around the far-distant
and just-discernible Sun, the great parent of them all.

[1] Pp. 485, 486.   [2] *Knowledge*, i. p. 124.

IX.

WE have now, after disposing of the Sun and stars, carefully, and with as much completeness as possible, considered the condition of the whole planetary and other bodies of the solar system, commencing with our Moon and ending with Neptune, the most distant known; and if others lie beyond Neptune yet to be discovered—and one has been suspected, and is the subject of diligent watch, but has not yet been apprehended and brought to give account of his errant life—their condition must be still more hopelessly inhospitable than that of any of the planets and their satellites whose case has been discussed. But it will be seen that, for reasons which seem to be unanswerable, each and every one of the planets—which (if the asteroids be reckoned, as there is no reason why they should not be) amount in number to thousands, to say nothing of the comets computed by millions—has been demonstrated to be in a state which, having regard to all the analogies of Earth, would render life upon them impossible; and this was the great problem which had to be solved.

It devolves on us, therefore, in closing, to review the results of inquiry, and form a conclusion. In doing so, it will be kept in recollection that after disposing of all the arguments which have been commonly used in support of the contention for 'plurality of worlds,'—such as the very familiar one, that the orbs around us would be useless unless inhabited,—certain positions were postulated as necessary to the just determination of the question in dealing with the several celestial bodies: because it would not do, for example, to say that as all things are possible with God, so God could give life to each orb in the Universe, no matter in what unfavourable circumstances it is placed. The results of investigation then are these:—

1. That life of any kind cannot be in the Sun and fixed stars.

2. That we cannot assume that all, or indeed any of the stars, as suns, have planetary systems, or that where such systems exist the planets constituting them are so constituted as to support vegetable and animal life.

3. That the Moon and all the planets other than the Earth, and all the other bodies of the solar system, are in conditions antagonistic to life.

4. That seeing the Earth is only one out of thousands regarding which we have means of forming an opinion, and that these thousands, we have good reason to hold, are uninhabitable; and also seeing that life has only

recently, by creative miracle, been bestowed upon Earth, —a world specially suited and prepared, if not specially made, for life,—we cannot reason because of life on Earth to life existing in any presumable stellar system.

The irresistible conclusion therefore is, that we can hold *life to exist only upon the Earth.* It may appear to be a strange and bold conclusion, but it is fairly rested on analogical deduction.

And with such a conclusion all the ideas drawn from the supposition of a plurality of worlds—pleasantly beguiling though they be—vanish and come to nought. All the fervent fascinating words, in which the inspiring glow of fancy has sometimes found utterance, and borne both speaker and hearer alike entranced away, are, like the sleeper's dream, rudely broken by the call to awake, stripped of power as the light of inquiry bursts in and dispels the delusion.

But have we no bright and noble thought to take their place? Nay, is it not rather in the contemplation to which this important deduction—this true ' creed of the philosopher '—leads, that there is found the more assuringly ' the hope of the Christian.'

We know that to the Omniscient Eye every celestial body, however large or however small, throughout the Universe, with all that passes on each, is as present as if one only of all were the subject of undivided oversight. We know, too, that the Arm of Omnipotence is, with unwearied effort, stretched over each to direct it and tend it with a solicitude as great and as unceasing as if each were the sole object of Divine regard. But when we find that this world alone, of all the worlds in space, is fitted for life, and has been prepared for the rational and immortal Being MAN, are we not justly led to take another and higher view of a planet, which has by some of its own denizens been slightingly designated 'insignificant,' and of the beings for whom it has presumably been called into existence, and harmoniously planned, and exactly arranged; and to feel all the more (at least in the weakness of our power to comprehend the Almighty) that there must be drawn to it and to them, in a manner peculiar and paternal, the never-ending concern and the boundless love of our heavenly Father. Can we then wonder that, conscious of His high designs, when He laid the foundations of the Earth, 'the morning stars sang together, and all the sons of God shouted for joy;' and when, after long ages of ministering preparation, the Earth was adjusted and furnished to receive a Being unlike, and so infinitely higher than, any which had previously subsisted on its face,—nay, as it seems, unique in the whole Universe; and God, seeing all was ready, had issued His gracious decree: 'Let Us make Man in Our Image,'—and had 'breathed into his nostrils the breath of life;' and, having planted a garden, had put therein the man whom He had formed, endued, though in a finite manner, with attributes resembling His own: can we longer wonder that He, thus regarding Man as His peculiar offspring, watched him with an anxious eye, and that His radiant messengers ever and anon sped their joyous way to reveal to this favoured being His beneficent will, His sovereign plans. Can we longer wonder that for his good there was bestowed on some a special power, and 'holy men of God' 'spake as they were moved by the Holy Ghost,' and gave to him an inspired Record to be the guide of his life, and to offer a hope for life to come. Or when this being, Man, the only rational, embodied, created spirit inhabiting the Universe, had sinned, fallen from his allegiance, rebelled, and forfeited his claim to all for which he was destined, can we not the better see why it was that the Eternal Son, filled with compassion, divinely infinite, for His fallen creature, withdrew for a time from the Majestic Glory in which He dwelt invisible and unapproachable, and, veiling that benign but dread face, upon which no mortal eye could gaze and live, He humbled Himself to take upon Him our nature, to put His foot upon our World, and for a time to associate in familiar intercourse with us, His erring children; and, having made known His mission of love and grace, submitted to taste of death that Man might, by that astounding sacrifice, be restored to the privileges he had lost; and then returning to His resplendent heavenly abode, to welcome there His redeemed people, He still retains for them a fellow-feeling—still watches over them with a jealous care—still listens to their every supplication; so that Heaven is, as it were, brought nigh to Earth; and, ever since the creation, our World, deemed to be so little among the thousands, has been sweeping on and on by the jasper walls, and almost grazing the jewelled sides; and ever as we passed the gates of pearl, which are never closed, we might imagine we heard the rustle of angelic wings, and that our eyes were opened, as were the eyes of the prophet's servant of old to see the horses of fire and the chariots of fire, and we beheld legions of bright shining ones, streaming out and flying quickly on errands of love and mercy to men, while 'the Ransomed of the Lord, with songs and everlasting joy upon their heads,' were as eagerly pressing in to march in triumph through the golden streets, lighted by the glory of God, and through the pure river of the water of life to join the innumerable throng which surrounds the Great White Throne. And if we can imagine Earth so linked to Heaven, and think of its inhabitants as so highly favoured and so highly prized, we are enabled the better to comprehend our lofty position, and we can proudly rejoice to think that though our world, amidst the millions of celestial bodies vastly larger, is—looked upon with a material eye—but an insignificant and indiscernible speck, yet, as not merely the sole dwelling-place of rational life, but the seat of the most amazing of all wonders, a being destined to a glorious immortality clothed for the present in a mortal frame, but conscious of his high prerogative; and therefore, and by reason of all that, because of its inhabitant Man, has supervened, it is a greater and grander globe than the largest and brightest orb which gleams upon the midnight sky.

CONCLUSION

upon with a material eye—but an insignificant, and indiscernible speck, yet, as not merely the sole dwelling-place of rational life, but the seat of the most amazing of all wonders, a being destined to a glorious immortality, clothed for the present in a mortal frame, but conscious of his high prerogative, and thenceforth, and by reason of all that, because of its inhabitant has, has superposed, it is a greater and grander globe than the largest and ruddiest orb which wheels upon the outer sky.

BY THE SAME AUTHOR.

In One Volume, post 8vo, pp. 490, with Twelve Lithographic Illustrations,
Second Edition, Price 7s. 6d. Cloth.

WINTERING IN THE RIVIERA;

WITH

NOTES OF TRAVEL IN ITALY AND FRANCE,

AND

PRACTICAL HINTS TO TRAVELLERS.

CONTENTS.

I. Continental Travelling—II. Hotel and Pension Life—III. Local Means of Conveyance—IV. Postal Arrangements—V. Sunday Abroad.

First Winter in the Riviera.

VI. London to South of France—VII. Cannes—VIII. Nice—IX. Mentone.

Italy.

X. San Remo and Genoa—XI. Spezia, Pisa, Sienna—XII. Rome—XIII.—Naples, Pompeii, Sorrento—XIV. Florence and Bologna—XV. Venice and Verona—XVI. Milan and the Italian Lakes.

Switzerland; France.

XVII. The Splugen Pass; Switzerland—XVIII. Biarritz—XIX. Pau—XX. Second Winter in the Riviera.

LIST OF ILLUSTRATIONS.

1. View from Foot-bridge up the Carrei to North Mountain Range, Mentone—2. The Estrelles from St. Honorat, Cannes—3. Oil-Mills, Carrei Valley, Mentone—4. Promenade du Midi, Mentone—5. Corsica, as occasionally seen before sunrise, Mentone—6. A City set upon a Hill, on road to Lucca—7. Sorrento from the West—8. Ponte Vecchio, Florence—9. Tomb of Juliet, Verona—10. Bellaggio, Lake Como—11. Port Vieux Bathing Establishment, Biarritz—12. Biarritz Bathers.

LONDON: LONGMANS & CO.

OPINIONS OF THE PRESS.

'Even to those who have not been, and do not intend to go, to the Riviera, the book will prove very interesting, for it is written in graceful English, and, where circumstances permit it, with no small descriptive power.'—*Bristol Times.*

'Mr. MILLER has fortunately remembered that there are still a few men and women left who have not been all over Europe, and he has not disdained to give particulars which some less considerate writers pretend to think unnecessary.'—*Manchester Guardian.*

'The Author of this record of an eighteen months' sojourn in the Riviera possesses the two prime qualifications requisite for the production of such record. He is a careful observer, and has descriptive powers of more than ordinary merit. . . . We think the present publication opportune, and certain to meet with acceptance.'—*Yorkshire Post.*

'Will prove eminently useful. . . . It contains a large amount of valuable information, and is written throughout with good taste and feeling.'—*Hull Packet.*

'There are some good sketches in the volume, and it is altogether one of the most interesting books of Continental travel that we have seen for some time '—*Aberdeen Journal.*

'The pleasantly told experience of visits made to such places in the south of France, Italy, and Switzerland as are often resorted to by English travellers in search of health or pleasure.'—*Ayr Advertiser.*

'For really valuable information, pointed and practical — just such as travellers, whether in quest of health or pleasure, desire to have before leaving home—nothing could surpass the volume under notice. . . . Will be widely read and enjoyed, not only by those about to travel on the Continent, but by stay-at-home travellers.'—*Daily Review (Edinburgh).*

'There is thus freshness and variety in the description of objects with which we are more or less familiar, and Mr. MILLER, who seems to have observed very closely the scenes through which he passed, is a very readable guide to the most entertaining portions of Southern Europe.'—*Newcastle Courant.*

'This book gives an excellent description of some of the favourite health resorts of the south of Europe, and is one highly to be recommended to intending travellers. . . . The various good illustrations give great additional interest to the volume.'—*The Ladies' Edinburgh Magazine.*

'There is much in it to instruct and amuse both those who have visited the same scenes as Mr. MILLER and those who hope to do so.'—*Scotsman.*

'Lastly, Mr. MILLER has the pen of a ready writer, and his style is easy and unpretentious. Thus his work is to be commended from all points of view, and we trust will meet with that degree of appreciation to which it is certainly entitled.'—*Land and Water.*

'At this season of the year, when many are contemplating leaving England for summer climates, "Wintering in the Riviera" will command attention, and doubtless suggest much that is of value to intending sojourners in the pleasant places it describes.'—*Morning Post.*

'Readers will be hard to please if they do not find such amusement and information as the prefatory pages modestly promise. . . . His observations of French, Swiss, and Italian scenes are all full of interest; and the faculty of accurate draughtsmanship and delicate sketching helps the Author to illustrate his descriptions, while it may not unreasonably be supposed to have reacted on his literary style, making it more concise, vivid, and exact.'—*Daily Telegraph.*

'Persons who contemplate to journey in that delightful locality cannot have a better guide and companion than the volume which Mr. MILLER has written for their benefit.'—*Newcastle Chronicle.*

'The Author of this work, writing in the interests of those who may wish to do as he has done, has hit the nail upon the head. . . . It says much for his work that we have read it all, and, though tolerably familiar with some of his scenery, with great satisfaction '—*Glasgow Herald.*

'Is a work to be classed with that recently published from the pen of Mrs. BRASSEY. . . . The volume is illustrated with several Lithographs, chiefly from sketches by the Author, and, as the views are very tastefully chosen and neatly finished, they add materially to the interest which every lover of travel must feel in the work.'—*Western Daily Press (Bristol).*

'Of everyday life at the hotels usually frequented by English visitors, the book conveys a good idea; some of the descriptions of scenery almost rise into eloquence; and although the Author cannot be implicitly trusted as a guide in matters of art, his remarks on the sights of Italian cities will be read with interest by those who purpose to spend a winter in the sunny South.'—*Academy.*

'The chief recommendations of the book are its impartiality and the opportunity it affords for a comparison of health resorts without studying a library of separate volumes. Mr. MILLER gives many facts which may be useful as to situation and scenery, exposure, climate, temperature, etc.'—*The Times.*

'To those who know Mr. MILLER as a man of culture and character we need say no more to indicate the sort of diary he has kept. For travellers of his own type—thoughtful, observant, and though not strongly prejudiced, yet thoroughly Scotch—he has written an excellent book. . . . It is pre-eminently a family book on winter resorts abroad. In that respect it is complete, reliable, and written with such good taste as to be agreeable and instructive reading.'—*Edinburgh Courant.*